Praise for *The Vulgar Wasp*

'Lester cleverly weaves facts and figures on the astonishing science of this little-loved insect into a text that's tickled with memorable anecdotes and personable insights. If you thought wasps were pointless, boring and unimportant, think again.'

—Dr Seirian Sumner, Reader in Behavioural Ecology at University College London

'*The Vulgar Wasp* is more than just interesting and instructive; it's a delight to read. It's packed full of up-to-date information on invasive wasps in New Zealand and elsewhere, as well as in their native range – all presented in an engaging, humorous and informative manner. Everything you might want to know about the life cycle of wasps, methods of control, and the environmental, social and economic costs of these pesky invertebrates can be found in this enormously readable book.'

—Dr Andrea Byrom, Director of New Zealand's Biological Heritage National Science Challenge

THE VULGAR WASP

The Story of a Ruthless Invader
and Ingenious Predator

PHIL LESTER

VICTORIA UNIVERSITY PRESS

VICTORIA UNIVERSITY PRESS
Victoria University of Wellington
PO Box 600 Wellington
vup.victoria.ac.nz

A catalogue record for this book is available from
the National Library of New Zealand.

ISBN 9781776561858

Printed in China by 1010 International

For my family

CONTENTS

Living by your wits is always knowing where the wasps are.
—Stephen King, *The Shining*

INTRODUCTION

Wasp!

One of my earliest memories of wasps is watching a hungry forager crawl in and out of my best friend's mouth. We were at school some four decades ago on a hot autumn afternoon in rural, small-town New Zealand. I've got absolutely no idea what the class was about. About all I remember from that class, and pretty much from the entire year, was that single wasp.

The wasp had flown in through an open window. She* ignored us other kids in the classroom. We were uninteresting. But a sweet smell from Chris's mouth was irresistible. We weren't supposed to eat in class; all the same, Chris was enjoying some sweet, fruit-flavoured chewing gum. The wasp hovered in front of his face for a second or two, confirming that here indeed was the source of the sweet odour. Those few seconds were just enough time for Chris to see the wasp and freeze. I'm still not sure if his response was one of fear or bravery – probably fear, but even now I suspect Chris would swear bravery. The wasp landed on his cheek. She then crawled over his chin, onto his lips, and onto his teeth. She came out again, crawled around his nostrils, then returned to his lips and teeth. For a short eternity, my teacher and classmates were silent and absolutely transfixed on this scene. In my mind's eye, the wasp in Chris's mouth was large, and she had several distinct black spots on the yellow sections of her thorax. (Later, I would learn that those spots and her large size would generally indicate a German wasp.) But everything was relatively big back then, and I'm not entirely sure that my imagination hasn't added the spots to my memory at some stage over the last 40 years.

---

* All wasp foragers – also known as workers – are female.

Chris's response was exactly the right one for this situation. He didn't jump, swat, swipe or even move a muscle. I suspect that he even stopped breathing and thus stopped the source of the sweet odour. His inaction demonstrated a lack of any threat to the wasp. The wasp finished her searching and flew off and eventually out the window. There were no swollen, throbbing fat lips or anaphylactic shock reactions in our class that day.

That memory highlights some key aspects of our relationship with wasps and their relationship with us. Chris's close encounter with a wasp is many people's version of a horror story. If you enjoy horror films (I don't, and I don't really understand people who do), you may have come across scenes in which wasps attack and kill people, crawling in and out of their mouth and ears, and laying eggs in their brains. The 2007 movie *Black Swarm* has yellowjacket wasps doing precisely that, only the wasps have been genetically modified to be used as military weapons. Of course, the wasps escape. And, of course, they turn people into zombies. What films like this really tell us is that we have an inordinate amount of fear and respect for these tiny creatures. We are around 800,000 times heavier than one of these insects. Yet I've seen people scream in terror at the sight of a single wasp worker flying by.

The common wasp (*Vespula vulgaris*). One of the distinguishing features between common wasps and German wasps is the black marking on the lower part of the face. Common wasps tend to have an 'anchor' shape. *Photo: George Novak*

More reasonable, however, is the level of fear and respect most people have for wasp *nests*. If wasps perceive a threat to their nest, they will react with a level of aggression that regularly lands people in hospitals – even morgues.

Chris's encounter tells us something about the wasp's perspective, too. She was just doing her thing: searching for food, maybe to feed others or to help her on her foraging trip – an insect's equivalent of a can of energy drink for the road. She was rather like a CEO entering a room full of her subordinates. She was without fear. She had a strong sense of her authority and the knowledge that this was her dominion. Her presence, suit and colouration immediately commanded respect. She didn't perceive a threat or feel the need to use her considerable weaponry. A degree of boldness, even aggression, but not too much. A lesson might have been needed if there was any hint of misconduct. But she modulated her behaviour for the best outcome for her and her family. It's hard not to admire that wasp.

The German wasp (*Vespula germanica*) generally has a stripe with distinct dots on its lower face. *Photo: George Novak*

Extraordinarily clever

Chris's wasp was either a common wasp (*Vespula vulgaris*) or a German wasp (*Vespula germanica*), both of which have been unintentionally introduced into New Zealand and into many other countries. The German wasp turned up in New Zealand after World War Two, thought to have arrived along with some plane parts being shipped here from Europe. And the common wasp became established in New Zealand in the 1970s. Wasps have typically arrived in their new countries as stowaways, often motionless ones. Overwintering queens have sought a dry spot to hide and sleep, such as inside a piano or piece of sawn timber, which leads to humans inadvertently exporting them all around the globe. North Americans would call these wasps and their relatives 'yellowjackets' because of their distinctive yellow and black colouration.

Wasps, bees and ants have been served well by their modulated boldness, aggression, and diverse range of behaviours that reflect a high level of intelligence. Biologists classify these insects as part of the order Hymenoptera. This name is probably derived from the ancient Greek terms *hymen*, meaning membrane, and *ptera*, meaning wings. It's also possible that the name comes from the Ancient Greek god of marriage, Hymen, because the forewings and hindwings of hymenopteran insects are often connected to each other by rows of little hooks: these hooks have 'married their wings' for flight. There are around 150,000 described hymenopteran species, with many more that scientists are yet to formally classify and name.

I've been fascinated by Hymenoptera for decades. For example, ants do much of what human beings do, but they've been doing it for a lot longer and a lot better than us. We 'invented' agriculture to grow crops, and we apply chemicals to these crops to control pests. But leafcutter ants have been growing fungus and applying antibiotics to cull unwanted species for millions and millions of years.[1] And they have been much more successful at it than we have been. The fungicides that we develop might work for only a few years before fungal resistance develops and we need new products. If leafcutter ants could, they'd scoff at our efforts! Other ant behaviours remind us of destructive, even immoral human behaviours – such as keeping others as slaves, and developing weapons of mass destruction. Even here, ants have developed their slavery and bomb-making abilities to a high level. For example, some slave-maker ants have evolved to be so reliant on slaves that they are totally incapable of rearing their offspring or even feeding themselves without their slaves' help.[2] A type of ant in Borneo even uses itself as a weapon. If the colony is under threat and in battle, the worker ants will squeeze their abdominal muscles so hard that they cause a mini-explosion. They and their enemies are coated with a toxic, sticky, lethal yellow substance.[3] This kamikaze tactic is the ultimate sacrifice for the good of their colony and the next generation. In 1884, the English scientist Sir John Lubbock wrote that

> the anthropoid apes no doubt approach nearer to man in body structure than do any other animals; but when we consider the habits of ants, their social organization, their large communities and elaborate habitations, their roadways, their possession of domestic animals, and even, in some cases, of slaves, it must be admitted that they have a fair claim to rank next to man in the scale of intelligence.[4]

The honey bee is the one hymenopteran insect that nearly the whole world loves. We are reliant on these bees for our food, though people have tended to overstate the importance of bees. Albert Einstein is often quoted as saying, 'If the bee disappeared off the surface of the globe then man would only have four years of life left.' Einstein almost certainly didn't say that (and given his lack of entomological and agricultural knowledge, I doubt that we could trust his opinion even if he did).[5] Another commonly cited statistic is that about one third of the human diet depends, directly or indirectly, on insect-pollinated plants.[6] This statistic is probably an overestimate, and perhaps only of relevance for more affluent economies, where hay-powered beef and dairy products, fruits and oilseeds make up a substantial fraction of people's diets.[7] Countries like New Zealand fall into that range of affluence, as reflected by my breakfast today of muesli, fruit, yoghurt, milk and coffee. All of these products are to some extent dependent on pollinators. The best estimate is that, while about 75% of crops benefit from animal pollination, only 10% of them fully depend on pollinators to produce the fruits and seeds that we eat. Bees and other pollinators collectively account for approximately 2% of our global agricultural production.[8] That amount is still impressive for one group of insects. It also appears to be growing.

Bees are clever little animals too. Their dance communication that indicates the direction, distance and quality of food sources is legendary. Their democratic election of new hive sites should be exemplary to many of governments around the world. Honey bees even self-medicate to combat pathogen infections: infected workers will choose and preferentially consume food with high antibiotic activity. Because these workers then feed the juvenile bees, the benefits of this medicinal behaviour are transferred to the entire hive.[9] Clearly, wasps have some very clever cousins.

When you think of wasps, you might think of a social wasp or yellowjacket, and you probably recollect the pain and indignity of being stung. But the vast majority of wasp species are of a parasitic variety, with a very antisocial nature. They are parasitoids. A parasitoid is an animal that spends a significant part of its life history feeding on a single host organism, ultimately killing that host. Parasitoid wasps find a caterpillar or other prey item on (or in) which to lay an egg. The egg then hatches and *eats its living prey alive*. The unlucky insect dies a slow, drawn-out death. The observation of these agonising deaths has been enough to turn believers into atheists. Charles Darwin famously wrote, in a letter to botanist Asa

'I am *not* happy about this!' A wasp worker with legs spread, sting extended and mandibles wide open expresses her dislike at this treatment. This wasp was part of an experiment to examine aggressiveness among individuals. Wasps from the same nest can vary in their behaviour. Upon receiving a mild electric shock, some wasps would not react at all; others struggled free and stung the researcher. *Photo: Davide Santoro*

Gray, 'I cannot persuade myself that a beneficent and omnipotent God would have designedly created the Ichneumonidae with the express intention of their feeding within the living bodies of caterpillars'.[10] The Ichneumonidae to which the great scientist was referring are one of the big families of parasitoid wasps.

Other parasitoid wasps include the smallest known flying insect, *Kikiki huna*, a species of fairyfly wasp. A full-grown adult measures a tiny 0.15 millimetres in length. The head and brain of *Kikiki huna* is a small fraction of that 0.15 millimetres, yet it still has the processing power to live a successful life that includes flying, avoiding predators, and finding mates, food and prey. Its closest known relatives are fairy wasps in the genus *Tinkerbella* (who says entomologists don't have a sense of humour?). Sawflies make up another big group of Hymenoptera. Sawflies are exclusively herbivores; they specialise in a diet of dead wood, leaves and other plant material. Some are major pests. The Sirex woodwasp, for example, is a major forestry pest. The female woodwasp lays eggs into living trees, along with a mucus-like substance and a fungus that kills the trees. Woodwasp larvae are only able to feed on dead or decaying trees, and would die in the absence of the fungus.

There are dozens of families of wasps. Only one of these families, the Vespidae, contains the social species that live together in nests. There are several different social wasp genera, including *Polistes*, or the paper wasps. In this book, however, I want to focus on the social wasps in the genus *Vespula* and, in particular, one species that has been ranked as one of the world's worst invasive species.

Vespula vulgaris: the vulgar wasp. The species name *vulgaris* is Latin for 'common', but 'vulgar' has connotations of behaviour that is coarse, rude, lacking in good taste or sophistication. *Vulgaris* is a frequently used species name and medical name. For instance, the starlings sitting in the trees outside my window are *Sturnus vulgaris*. These are common birds native to Europe and western Asia; they are now even more common, having been introduced to far-flung countries like the Falkland Islands and Uruguay.* There are no points for guessing what type of animal *Octopus vulgaris* is, nor for guessing that it is widespread. And, like many of us, you may have been intimately familiar with *Acne vulgaris* as an adolescent.

---

* Precisely because starlings were mentioned once in Shakespeare's *Henry IV*, 100 of them were released in New York's Central Park in 1890. They have since become common in every state in the US.

Vespula is the genus name. It's derived from the sister genus *Vespa*, which are hornets, and is Italian for 'wasp'. *Vespula* are phylogenetically differentiated from their sister genera by traits such as a relatively short space between their eyes and mandibles. And this genus has a tendency to nest underground. Perhaps more than 95% of the nests we come across are underground, often with only a few inches of soil covering them, which people discover when they step into them.

These wasps are frequently and quite literally a pain in the butt. But a part of me admires them too. These little insects with tiny brains can be extremely clever. Their nest architecture is stunning – the colours and patterns of their nests are even emulated by Japanese artists. Their stinging behaviour is not as mean-spirited as you might think. How would you feel if a monster, which you would immediately assume to be a predator, put its hand or foot through your roof and into your home? Would you react passively, or would you use everything in your power to persuade the monster to go away? Wouldn't you do your best to teach it never again to attempt to chow down on your family? These wasps

A German wasp worker foraging. *Photo: George Novak*

have evolved with animals such as badgers and bears, which do very much like to snack on baby wasps. For all we know, when we stumble on their nests they assume we are about to eat them.

The invaders

The vulgar wasp is known as a biological invader. Dan Simberloff, one of the leading scientists in this field, defines a biological invasion as the human-assisted arrival of a species into a region where they are not native, and that species' subsequent establishment of a population. 'If the population then spreads in its new home,' he writes, 'the phenomenon is called a biological invasion and the species is invasive, at least in this region.'[11] The human-assisted movement can be intentional or unintentional. Common and German wasps in Australia, South Africa, South America and New Zealand certainly fit these criteria. Both species arrived with unintentional human assistance by stowing away on human cargo; they then became established, and spread. Many other scientists and management groups define biological invaders using these criteria, with the addition that the species is causing some sort of negative impact. The International Union for Conservation of Nature defines biological invaders as an 'invasive alien species'. This is a species 'established outside of its natural past or present distribution, whose introduction and/or spread threaten biological diversity'.[12] The IUCN recognise that not all alien species have negative impacts, and estimate that between 5% and 20% of all alien species become problematic.

This is the ongoing story of an invasive species. The introduction, adaptation and acceptance of many invasive species will continue to contribute to our current biodiversity crisis. We are in the midst of what many scientists describe as the sixth mass extinction event.[13] Extinction is known as a natural phenomenon – we can find evidence for species becoming extinct in our geological records reaching back hundreds of millions of years. But, once every few hundred million years, we see evidence of extraordinarily high extinction rates. The previous five mass extinctions in the world's history have occurred due to a range of events, including huge meteor strikes and massive volcanic eruptions. But right now we are seeing extinction rates that are estimated at 1,000–10,000 times the normal rate.

Mexican researcher Gerardo Ceballos and his co-authors have described the globe's biodiversity crisis as 'a massive anthropogenic erosion of biodiversity and of the ecosystem services essential to civilization. This "biological annihilation" underlines the seriousness for humanity of Earth's ongoing sixth mass extinction event'.[14] The 'anthropogenic erosion' is due to factors such as habitat loss, overexploitation, pollution, climate disruption, and the effects of invasive and introduced species. Introduced predatory mammals – such as the Indian mongoose in the Pacific Islands, for example – have resulted in the local extinction of native birds, reptiles and amphibians. Brown tree snakes in the island nation of Guam have similarly devastated their bird and reptile fauna.

Indian mongoose, brown tree snakes and our common wasps are all listed in *One Hundred of the World's Worst Invasive Species*, a publication that seeks to illustrate 'the incredible variety of species that have the ability, not just to travel in ingenious ways, but also to establish, thrive and dominate in new places'.[15] The 100 least-wanted species were chosen subjectively by ecologists and biologists who have been working on exotic species incursions for decades. They made their choices based on the seriousness of each invader's impact on biodiversity and/or human activity, and as representatives of important issues associated with biological invasions. Of the 100 invasive plants, animals and microorganisms, there are six social insect species. That vulgar or common wasps and five other social insects feature here speaks volumes about the success and impact of these animals. Our accidental or sometimes intentional movement of species around the globe is frequently ranked second only to habitat loss as a cause of species extinction and endangerment.[16] This book is about invasive wasps, with a focus on *Vespula vulgaris*; this book also reflects the situation for those other 99 invasive species, and many more. Many similar invaders have human and environmental effects, with considerable debate and discussion on what to do about them.

The scientific discipline of invasion biology is much more contentious than many might realise. In his book *Where Do Camels Belong? The Story and Science of Invasive Species*, ecologist Ken Thompson warns us against trying to force our biodiversity to return to the 'frozen moment' of some point in the pre-human, pre-industrial past. That 'frozen moment' is a common goal around the world –

indeed, one wildlife sanctuary in my home town of Wellington has that specific goal and purpose. Ken argues that, in the United Kingdom and elsewhere,

> successful species, alien or native, are symptoms of change rather than drivers of that change, and all they are telling us is that they are very pleased with the changes we have made to their environment. We have the plants (and animals) we deserve.[17]

Ken believes it is a widespread myth that 'alien invasions reduce biodiversity and ecosystem function' and that 'aliens are bad, natives good'. He paints a dim view of scientists who, like myself, focus their research on invasive species. People like me, who study biological invaders, are apparently guilty of seeking funding, a full research laboratory, and fame.

To be fair, Ken acknowledges that there are invasive species that can be easily classified as 'a bad thing'. Guam's brown tree snake invasion is one such example; he believes that efforts to control these and some other invaders are appropriate. He concludes by saying that we should carefully consider how we manage invaders: 'I am saying that we should commence any attempt to control alien species with our eyes wide open. We should be certain, from an honest, objective analysis of the best available evidence of its positive and negative impacts, that our intended target is causing *net* harm.'[18]

Here, I agree with Ken. Wasps are dramatically altering our biodiversity. But they are doing some good, too, such as through their involvement in pest control and perhaps some pollination. Do *Vespula* wasps exert '*net* harm' in their invaded range? Calculating the net effect of an invasive species is not without its challenges, because people's value systems vary. In the eyes of some, an endangered butterfly or bird has high value and any introduced predator that threatens those endangered species thus causes substantial harm. Others might not care about conservation, only becoming concerned if the endangered species or the introduced predator were to influence their bank balance. Whose value system should we use to calculate the effects? How do you incorporate financial, social and conservation effects into your calculations?

Should we manage invasive social wasps, or just accept their presence as part of our new world? If we choose not to adapt to their presence, what should we do to control them? Would people accept poisons repeatedly raining down across landscapes from helicopters? Or, should we use new technologies to eradicate wasps from entire countries? Potentially game-changing technologies are being developed and debated right now. Gene drives that rapidly spread mutations or deleterious genetic modifications have real potential to achieve widespread eradication of many pests and diseases. In another 10 years, it is possible that this technology will be ready to eradicate pests from whole continents. For example, several laboratories are developing gene drive systems to control mosquitoes and the malaria that they carry.[19] Should we use gene drives on *Vespula* wasps to exterminate every last individual from Australia, New Zealand, South Africa, and anywhere else we humans decide that we don't want them to be? How do we go about getting a 'social licence to operate', effectively involving everybody in these decisions?

The wasp is a maligned and misunderstood creature. Unlike the bee, it is near universally hated. In New Zealand, Britain, Switzerland and throughout their native range, social wasps are an annual target for pest controllers and do-it-yourself destroyers armed with bottles of petrol and matches. Nearly every year, people die, suffer burns, and are even made homeless by attempts to kill wasps. Arguably, human beings have made the situation worse by giving some social wasp species unfortunate names, which can influence the way we behave around them.

One of my many hopes for this book was to translate a lot of dense scientific information for anyone who has a passing interest in wasps. Much of the literature on wasps is pretty old. It spans the last century and longer, representing entire careers of dedicated scientists. Their work is admirable in depth and still very relevant, but it is usually locked up behind pay-per-view scientific journals or in dusty old books that are rarely checked out of libraries. This is a shame, because common wasps and their relatives have a fascinating life history. Though we know them as pests, they are amazingly efficient predators with some exceptionally smart behaviours. The common wasp excels as a hunter and an invader.

In writing this book, I want to highlight the many sides to wasps and pests. I hope to present our options for the management of wasps in a way that is comprehensive and unbiased. Most of all, I hope to tell the story of the vulgar wasp and its impact on us and on the world's biodiversity.

Wasps ahead. *Photo: Jacqueline Beggs*

1. WHEN THE BABIES FEED THE PARENTS

A wonderfully weird life history

Which comes first: the wasp or the wasp egg? To describe the life cycle of a wasp colony, let's start with the overwintering of foundress queens. These are the new wasp queens produced by a nest at the end of autumn. They, and the good-for-only-one-thing males, are the goal of the entire spring, summer and autumn's work by several generations of female workers. Most of the workers born to that nest have given their entire life's work and have sacrificed their reproductive ability without ever seeing these new queens and males that represent the ultimate fruits of their labours and lives. After these foundress queens leave the nest in autumn, the old queen and remaining workers die. The colony disintegrates and hums no more.

Overwintering

For social wasps, overwintering – or hibernating – is not a social event. The new foundress queens spend the winter alone or occasionally in small aggregations under the bark of trees, or anywhere else dry and safe from predators who would otherwise enjoy a docile and fatty meal. The most attractive hibernation environments are all too often human-made ones – such as log piles, pianos, sheds and shipping containers. Unfortunately, these are the kinds of habitats that are also sometimes sent overseas. A quarantine or biosecurity official will find it hard to spot a solitary hibernating wasp queen tucked away in a tight corner of a shipping container. It's likely that the queens that resulted in common wasps establishing themselves in New Zealand in the 1970s were overwintering queens, shipped from Western Europe or Britain.[1] Entire nests, on the other hand, with their active and stinging workers, are much more easily observed (or felt) by people attempting to ship the goods and by quarantine authorities on the receiving end.

New wasp queens produced in autumn nests have body fat of up to 40%, so would be at least substantially overweight by polite human comparison. Thinner and leaner queens are also produced in autumn, but they're poorly represented among the surviving spring populations.[2] To survive the winter, the queens rely on enlarged fat cells and a dramatically reduced metabolism. Other social wasp species have additional tricks to help survive cold winters, such as with queens in the related genus *Dolichovespula*. These wasps have ice-nucleating agents that promote freezing outside the cells and prevent the formation of ice inside vital

A common wasp queen overwintering. In autumn, new queens mate with one or more males. The queens are then able to store the sperm over the winter. If a queen survives the winter, she emerges in early spring to start a nest. *Photo: Dave Hansford*

body cells.[3] Effectively these wasps have a type of antifreeze in their bodies. But despite these clever adaptations to the cold, one estimate puts the overwintering survival rate of wasp queens at just 2.2%.[4] For every 100 new queens produced, only two will survive the winter and establish a nest the following year. That's a high mortality rate. But, given that an individual wasp nest can produce hundreds or even a thousand new queens, in the absence of such mortality we would see many, many more wasps around us.

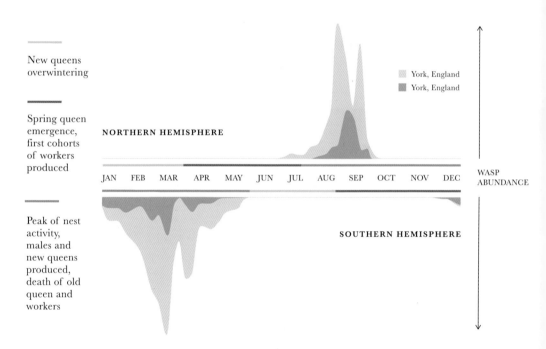

Timeline showing the population dynamics of common wasps in an invaded range (Lake Rotoroa, NZ) and in a native range (York, England). The timeline is broken into three phases to represent the wasp's life cycle. In England, a year of relatively low wasp abundance was followed by a year of higher densities. The differences in abundance for the NZ data were achieved by pesticides. The milder climate results in a much longer wasp season of around six months, compared with four in England.

Obesity is clearly an advantage if you aren't going to be eating for months on end. Several scientists have hypothesised that when there is an abundance of these well-provisioned (obese) wasp queens that survive overwintering, they have a major influence on wasp population dynamics.[5] In Britain, 'wasp years' are often characterised by few wasps in the preceding year, followed by a season with wasp numbers often hundreds of times higher. This type of scenario is known as density dependence: the density of individuals in one generation has a big effect on the following generation or generations. The single biggest predictor of wasp abundance in any one year is the density of wasps in the previous year.[6] One potential explanation for wasp years is that, in seasons with low density, the absence of competition between wasp nests allows great numbers of fat queens to be produced. These 'high-quality' or healthily fat queens tend to survive winter, and in the following year we see high densities of wasps.

Flight and fight

It's spring. While you are celebrating the tulips and daffodils, foundress wasp queens are waking up. An 1868 book on wasps by amateur entomologist Edward Latham Ormerod describes this scenario in England: 'With the first promise of spring, with the violet and the primrose, with the snake and the bee, on the same bank, from which the warmth of the sun has called them all forth, the mother wasp enters on active life.'[7] This romantic evocation is in stark contrast to how most New Zealanders feel about their first sight of wasps in the spring and summer.

These wasp queens are frantically busy. First, a queen will fly away from her overwintering home in search of a nesting site. Most queens disperse some distance away from their winter homes, and are good fliers. Research in Argentina suggests wasp queens can cover very long distances: the rate of spread of the invasive German wasps was 37 kilometres per year (though it's likely that people help wasps to go that distance). If you were to tie a German wasp queen to a 6-centimetre-long stick on a pivot and have her fly around in little circles for as long as she could, she might fly further than 5 kilometres before giving up.[8] Perhaps the wasp is dizzy by that stage. Or really bored. Either way, that's an awful lot of little circles. The range of flight distances observed in the Argentine study correspond well with observed dispersal by German wasps in New Zealand. Related wasps can fly immense distances, sometimes in immense numbers, but a German wasp emerging in spring probably flies only a few hundred metres to find a new nest site.

One extraordinary dispersal or migration event was documented in 1957 by Swedish ornithologist Gustaf Rudebek. As he wrote in his field observations, while attempting to research bird migration he (understandably) found wasps much more interesting. He and his friends had 'most eventful' days counting queens flying from Sweden out to sea towards Denmark, where the closest landing point was 25 kilometres away. Gustaf reported peak migrations of more than 80 wasp queens per minute, flying in a stream 20 metres wide and 0.5–3 metres up from the ground.[9] We can extrapolate from his data that potentially 36,000 wasps each day were abandoning Sweden for the greener pastures of Denmark.

In a later study, an even more impressive migration was observed by Finnish entomologist Kauri Mikkola. He was on a ship sailing from Estonia to Finland, and observed wasps and bumble bees flying the 80-kilometre stretch across the gulf. Kauri also observed a flight of 200,000 wasp queens migrating in an easterly direction past Helsinki over just a three-day period. He deduced that the whole of southwest Finland contributed to this refugee wasp population.[10]

Why do wasps in Scandinavia appear in such masses to migrate such long distances? The short answer is we don't know, though one hypothesis is that weasels are to blame. Weasels hunt voles, and over several years we can see 'boom and bust' dynamics in the population cycles of voles and weasels – that is, the steep rise of a population is followed by a dramatic fall. Years with large numbers of voles lead to years of high numbers of weasels. It's during the big weasel years that ground-nesting social insects like bumble bees and wasps suffer. When a weasel is searching for voles in holes in the ground, a newly founded and poorly defended wasp nest represents a tasty intermission snack. To escape all of these weasels, wasp queens need to fly many, many kilometres. Hence their long flight from Finland over water.[11]

Old rodent burrows are highly sought-after as nesting real estate for many *Vespula* species, including common wasps. They will have used up most of their reserves in the overwintering period, but then they have to dig or excavate a new home. Crevasses and cracks in the ground or in fallen logs are highly prized. And once she has initiated a small nest and a cohort of workers, the queen must go out foraging to feed her brood. That a small mammal can get its weaselly teeth around a queen and devour her incipient colony, and perhaps even evolutionarily shape the migration behaviour of wasps, highlights the extreme vulnerability of these young nests. During this dangerous time, when a nest is only just beginning, the considerable benefits of being a social species are absent. Social insects are thought to be so abundant and frequently dominant because tasks like gathering food and defending the nest can be partitioned amongst colony members. Overlapping generations contribute to the diversity of necessary tasks that include nest defence. The workers sacrifice their lives and ability to reproduce to help raise the brood and ensure the success of the colony as a whole. But in early spring, when the nest is young, these workers and their defensive, stinging and sacrificial nature are absent.

While weasels and other small predators can be an issue for our new queens, perhaps the biggest threat at this time is other wasp queens. Young nests

A common wasp builds her nest. Wasps collect wood from different trees and other sources, which they mix with saliva and paste in strips, giving the nest a beautiful layered effect. (This wasp has a Radio Frequency Identifier Tag on her back so that we could monitor her foraging behaviour.) *Photo: Phil Lester*

frequently show signs of violent, gladiatorial battles to the death, with carcasses of usurped and defeated queens from previous fights rotting in the basement, so to speak. Researchers who have dug up and dissected early-season wasp nests have observed that more than 80% have been started by one species and then taken over by another.[12] A queen might initiate a small nest and produce a small number of workers – but then a queen from a different species arrives. Any wasp workers that attempt to protect their mother and queen are immediately beheaded or cut in two: workers stand almost no chance in their attack against the considerably larger and heavily defended queen usurper. Eventually, the two queens meet. A battle ensues: each queen must find a vulnerable spot on her opponent's body to deliver a fatal dose of venom. The 'winner', however, might not truly win. Injuries sustained in battle can prove lethal over the following hours or days. But if the new, usurping queen wins the battle and survives, she will take over the nest and have the existing or emerging workers do her bidding – at least, for a short time. The winning queen should expect subsequent corporate takeover attempts. Some nests have been observed with six queens lying dead and defeated from usurpation fights on the nest floor.

Ecologists refer to this type of interaction as social parasitism. The extreme social parasites of the insect world are the slave-making ants, which are morphologically and behaviourally unable to carry out such tasks as rearing their own young. These ants rely on finding and raiding other ant species to recruit slaves for these and other menial tasks around the colony. The slaves will even join additional raids for other colonies, aiding in the pillaging of pupal ants for future generations of slaves. But, just as in human slavery throughout history, the subjugation and submission of slave ants isn't necessarily total: the slaves can rebel. The slaves can successfully nurture, tend and produce pupae from their own species, while tearing to bits and killing a third of the pupae of their master. Or, if they don't tear them to shreds, the slaves might simply carry these ant pupae outside the nest and leave them there to waste away.

Within our social wasp genus, one of the extreme parasites is the cuckoo yellowjacket wasp, *Vespula austriaca*.[13] The cuckoo wasp has a longer hibernation period than other wasps; it only wakes up from its nice sleep-in when its potential hosts have built small nests. It forages for a drink of sweet nectar before starting its search. Once our cuckoo wasp finds a host nest, it invades. This is the *modus operandi* of the cuckoo wasp: instead of producing workers of its own, it parasitises other *Vespula* wasp nests. Initial interactions between the host and the cuckoo queen have been described as relatively calm, even civil. When aggression is finally initiated, the intruder will battle and kill the host queen, often with a powerful sting to the head. The cuckoo wasp is built tough for winning these fights, with wide mandibles, extra muscle and dense armour. It will chase down workers and maul them until they regurgitate food. After the cuckoo wasp lays its eggs, the host workers are forced to rear its larvae.

This is a fascinating interaction, even more so because much of the behaviour appears chemically mediated. It's completely dark in the nest, so vision isn't an option. The cuckoo wasp has larger than average pheromone glands that it uses to subtly, delicately manipulate its host. For example, these pheromones will suppress the reproductive development and behaviour of its host. I think this is amazing: here is one species using chemical communication signals to manipulate the behaviour and physiology of a completely different species.

This first phase of nest site selection and nest building is critical for a queen's colony. If the queen and her incipient colony survive usurpers, cuckoo wasps, and rodents with sharp teeth, there is a good chance that the nest will succeed.

The life and times of wasp larvae

The building of wasp nests starts with an attachment to the ceiling of the underground cavity. A small tree root will do nicely. This initial pillar may be only a few millimetres thick but is strong enough to support a nest that might grow to more than 10,000 cells in just a few months.

Each nest is constructed entirely of paper that the wasps make themselves. Wasps scrape wood from trees and even fence posts, and mix it with their saliva to form a pulp. When a little ball of wood has been collected, about the size of a wasp's head, the worker will fly back to the nest. Different wasp species have different wood preferences – German wasps collect pulp that produces a grey nest; common wasps typically use rotten logs, which produce a pale brown or yellowish nest. Since different wasps collect pulp from different wood sources, beautiful patterns can form in the nest, especially those of common wasps.

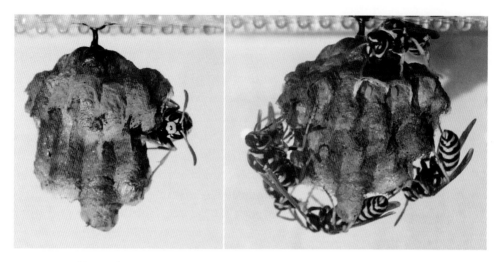

Nests of the European paper wasp (*Polistes dominula*) are usually grey, but here some captive colonies have been given coloured paper as nesting material.
Photo: Mattia Menchetti

Potters in Japan have admired and attempted to mirror these patterns in their clay. Even more fun than wet clay, if you have way too much time on your hands, is rearing paper wasps in the *Polistes* genus. Provide them with different-coloured papers, and eventually you'll have a rainbow paper nest. This sort of behaviour can keep graduate students amused for weeks on end.

Much as in the hives of honey bees, wasp nest cells are hexagonal in shape and sit side by side in a layer of comb. But, in contrast to bees, wasp workers build these hexagonal cells vertically. The eggs and larvae hang upside down like bats in a cave. Workers keep the size of the cells consistent by using their antennae like measuring sticks. The flat, symmetrical, hexagonally patterned comb layers are especially impressive when you remember that they are built in total darkness. To make extra layers of comb, the wasps build pillars that connect to the upper comb. A typical nest will end up with many layers of comb, like a stack of pancakes. Each flat 'pancake' is separated by the height of a wasp queen (less than a centimetre). At the end of the wasp season, the workers build comb layers at the base of the nest. These layers have larger cells, specifically for queen rearing. Layers of thin 'envelope' surround the nest and are built and extended with each new comb layer, with an entrance often at the lowest point of the nest.

While the biggest predictor for the abundance of wasps in any given year is their abundance in the previous year, the next most important factor is the spring weather. Rain and cold are not friends to the growing wasp nest. If you live in a basement, you'll probably sympathise with wasps, who must get a little concerned by the pitter-patter of rain on their soil roof and the resulting water running through their nest entrance, flooding it from the ground up. If your only way out is at the bottom, the water flow is a problem. During wet springs, many subterranean wasps must be trapped and drowned by a rising rain tide.

In early spring, a wasp queen is anxious to grow her colony. She lays her eggs even in cells that the workers haven't finished building. Her eggs are coated with a sticky adhesive chemical that glues them in place. And after around five days, the eggs hatch. The outer shell of an egg remains crucial for a wasp larva over the next several days, as the larva anchors itself to the cell by leaving some of its body segments within that broken shell. If at any stage of its juvenile life the larva

Wasp workers tend larvae. The layer of comb contains mature larvae and pupal cells, capped with a fine layer of silk. This nest would normally be underground and in complete darkness. The greyish, hairy wasp at the top left has just emerged from her pupal cell. *Photo: Dave Hansford*

drops from the cell, it cannot be rehoused by wasp workers. Instead, workers pick up any fallen larvae and dump them outside the nest, as refuse. This behaviour has probably evolved because it's usually only the diseased or sick larvae that fall. For the sake of the colony, it's best that any diseased individuals are well away from their young sisters. Successful larvae stay glued to the top of their cell.

All insects have an exoskeleton, which limits their growth and expansion. The exoskeleton of an adult is hard; it is the insect's armour. An adult doesn't grow or change its exoskeleton, so whatever size it is when it emerges from its pupal cell is the size it will be for its entire adult life. Larval wasps have an exoskeleton too, which keeps their organs and body fluids in place, though it is much thinner. Plus, kids will be kids and they need to grow. Wasps have five distinct stages of growth. In insects, these stages are known as 'instars', or moults. At the conclusion of each instar, a new, larger exoskeleton is formed underneath the old. The old one splits and the larva wriggles out to eat and expand into its larger form.

These wasp larvae exhibit some strange behaviours. For the first three instars, the larva will face outwards, away from the centre of the nest. For the fourth and fifth instar, the larva will leave the remnants of its egg shell and face inwards. We have no idea why the larva decides to turn around. As the nest is completely dark, it seems unlikely to be for a better view of the scenery. The larva remains head down. At this stage it is large enough to keep its grip on the cell using its lobes or ridges, with a grip strong enough to resist a researcher attempting to pull one out with forceps. (It is not uncommon for us to accidentally pull the head off a larva rather than remove its entire body from a cell.)

Wasp workers will access and provision the larval cohort, collectively known as the brood, from the sides of the nest. They attend to a layer of larvae by crawling over the top of the comb ceiling directly beneath them. To get attention from passing adults, the larvae violently shake their heads and snap their mandibles, making a loud rasping noise on the sides of the nest cells. This is the hunger signal, and it's loud enough to be heard unaided by the human ear. A worker will visit, inspect each larva with its antennae, and provide a drop of liquid or a semi-chewed mouthful of insect prey or meat. But this is not a one-way street. In return, the larva will secrete a droplet of liquid, which the worker swallows. The secretion is high in sugars such as glucose, and contains some protein. One researcher calculated that a worker can survive for about half a day after eating a

A common wasp queen tends a mature wasp larvae. (This picture is upside down, as larvae hang upside down in the nest.) Larva receive food from the queen or the workers, and provide a sweet secretion in return. *Photo: Dave Hansford*

single droplet from a late-instar larva.[14] The workers enjoy this sugary secretion, but it does come at a cost to the larva. If too much is asked of too many larvae, a colony can shrivel and die. And in another wholesome exchange, larvae also help digest protein meals for the adults. The adults provide some chewed-up food to the larvae, then the larvae add digestive enzymes and regurgitate a semi-digested, protein-rich liquid back to the adults.

A particularly weird behaviour of wasps in their larval stages is that they don't poo. Instead, a larva stores undigested waste products from its food in its mid-gut. Its final few body segments, which (as per usual for insects) contain the anus, are wedged into the remains of the egg shell. That wedging precludes any bathroom action for the larva. If it were to lose control and go for a poo, it would likely drop from its cell to be unceremoniously removed from the nest and dumped outside by an adult worker. So it's best to hang on. The kidneys (known as Malpighian tubules in insects) are present, but researchers have suggested that they have no function at this stage.[15] By the late fifth instar, the larva is a white, translucent, legless form with a large, semi-solid black mass of faeces in its middle. This faecal mass might weigh as much as a third of the larva's body weight and is carried through all the instars of the insect. The larvae void their faeces only at the start of the pupal stage: an action which must be of considerable relief for the severely constipated insect. The faeces effectively form a cup at the base of the cell (which you'll remember is actually at the top, because of our upside-down juvenile wasp). The larva then spins a complete cocoon that encases the pupa separately from its faeces and the cell. The entire larval life lasts for approximately 9–11 days. The whole process to produce an adult common wasp takes 23–29 days. Smaller, early-season nests produce smaller workers faster, while older nests produce larger workers over longer durations.

Wasp larvae are tasty to predators (and people)

A common wasp emerges from a pupal cell. The long antennae indicate it is a male wasp. Male wasps do nothing to help the nest – they don't collect food or have stings to help defend the nest. *Photo: Dave Hansford*

It's when wasps are in their older larval and pupal stages that they are most desired by animal predators. But the stings of the adult wasps appear to sway the mind and action of all but the most determined. In Europe, badgers are some of the most aggressive predators. Being repeatedly stung by wasps probably makes your average badger even angrier and more determined.

And then there is the honey buzzard – one of the very few specialist vertebrate predators of wasps. This bird's breeding season is timed for wasp and bee abundance, which tells us just how important these insects are to the bird's life history. This buzzard, despite its name, isn't a buzzard. It's a kite, a bird of prey that is classified by having a weaker bill and feebler talons than the buzzards. This is a special bird, however you want to describe it. Honey buzzards will catch and kill some prey that is only mildly tasty. But this is a calculated move. The honey buzzard doesn't eat the whole carcass. Instead, it leaves a small portion of meat out as bait for wasps. A foraging adult wasp will find the meat, cut off a little and fly back to its nest, closely followed by the honey buzzard. The Judas-wasp effectively leads the buzzard back to its home.

In Europe, common wasps nesting underground are a preferred prey for the honey buzzard. Its attack begins with the wasp larvae and pupae as the prize. The bird uses its armoured, heavily scaled talons to dig and rip open the nest envelope. Its face has scale-like feathers for protection, and it also has a substance on its feathers that is thought to repel wasps or deter them from retaliating. The feathers are tightly laced to make penetration difficult for the wasps. Some honey buzzards even work in teams to rebuff the wasp attacks. In Japan, honey buzzards will successfully take on nests of the Asian giant hornet – a 4-centimetre-long insect that is responsible for dozens of human deaths and thousands of hospitalisations each year in Asia. We often hear about chimps and crows using tools cleverly, but the buzzard's use of meat to bait and find wasp nests is an exceptional use of tools as well. Most other vertebrate predators sensibly stay well away from large wasp and hornet nests.

Buzzards and badgers seem to be on to something, as it turns out that wasp larvae are actually pretty tasty to some of us. Although the idea of eating insects is largely shunned by European cultures, in many other cultures insects are a delicacy and a treat. Particularly in parts of Japan, wasps are highly valued and sought-after — a caviar of the forest. The demand for them appears to exceed supply.[16] In the past New Zealand has exported wasp larvae to Japan, which made a nice change from exporting sheep.

The documented history of insect consumption, or entomophagy, in Japan dates back to at least the 1600s, though it is likely to have occurred for thousands of years. As well as eating wasps, the Japanese have found ways to enjoy mayflies, grasshoppers, dragonflies, hornets and beetles. The consumption of insects is a tradition in some central and rural areas. The Kushihara Hebo Wasp Festival, for example, annually celebrates wasp-eating and even has a prize for the person who presents the largest wasp nest. (The 2013 prize was won by a nest weighing 5.04 kilograms. The size and success of the winning nest might have been due to 'the poisonous snake that I lovingly killed and fed to [the wasps] just a month

Entomologist Bob Brown (in full reinforced bee suit) removing a common wasp nest from the ground. Wasps don't appreciate having their nest disturbed and will vigorously defend it. *Photo: Dave Hansford*

before'.)[17] European commentators and even some Japanese have sometimes described insects as 'famine food'. But this characterisation doesn't seem accurate. For example, Emperor Hirohito (1901–1989) described wasps as his favourite food. He ate 'wasp-rice' even when he was ill and unable to eat anything else.[18]

Charlotte Payne, a researcher from the University of Cambridge who focuses on edible insects, has tasted many different insects around the globe and rates wasp and hornet larvae among the most delicious. Her work surveying insect preferences in Japan also suggests that adult wasps are particularly tasty.[19] One enthusiast, after trying wasps from Kushihara, blogged: 'The wasps were small, soft and black: perfect for taking pinches of three or four in your hands and washing them down with some homemade wine. Some were savoury, others were sweet – it was almost like salty-sweet popcorn.'[20] Of the 117 insect species known to be eaten in Japan, wasps are eaten by people in 42 prefectures and are the most preferred by nearly all age groups. A popular dish is tsukudani, in which wasp larvae are simmered in soy sauce and sweet rice wine (mirin). You can also have your larvae salted and steamed, or baked into biscuits made with rice flour. Adult wasps can be made into cookies too. Large adult hornets are (more carefully) collected alive and are added directly to a distilled liquor similar to vodka – the hornet liquor is thought to have medicinal properties. Dishes and delicacies involving wasp larvae have received rave reviews. Hornet liquor, on the other hand, receives a somewhat lower rating. 'Let's be frank,' says one reviewer. 'It stinks, it tastes bad, and it looks like crap. I can't say that I would ever want to drink this stuff again.'[21] However, in Japan, like nearly every country today, attitudes and behaviour are changing. Many Japanese wouldn't consider eating insects. Rural residents in many prefectures might have eaten a diverse range of insects 30 years ago, including aquatic species and grasshoppers. Now, a more common attitude is, 'I can eat wasp larvae, but not any other insect'.[22]

There are many ways people can collect wasp nests. One way is to use bait, like the honey buzzard does. Wasps like to eat meat, so one practice is to cut up a dead animal such as a frog into tiny pieces and tie a silk strand to each portion. These silk strands are light and easy to see. When a wasp picks up a Kermit nugget and flies back to its nest, you can chase it by following the strand of silk. If you are quick enough, you will find the nest. Then you can place a firecracker in the nest entrance so that the smoke paralyses or confuses the wasps, before you starting digging. Charlotte, who spent time collecting wasps in rural Japan, told me that collecting a nest can also be as simple as banging a shovel or stick on the ground, after which the adult wasps go back into their nest. The Japanese *Vespula* species are quite placid insects, and their stings apparently 'hardly hurt at all'. But in New Zealand, when we bang a shovel on the ground near a common wasp nest, typically wasps flood out of the ground with a clearly murderous intent. Any gaps in our protective suits are quickly discovered and penetrated. What follows is your research assistant running off as fast as their legs will carry them to tear off the suit in a desperate attempt to dispatch the stinging wasps. The supervisor during this process is busy trying to stifle his laughter. In stark contrast, in Japan, early season nests can be easily taken away and housed in special boxes. Later-season nests are harvested for food, or, after discovery, might even be left as a source of queens for the following year.

However, traditions of hunting and keeping wasp nests appear to be dying in Japan. These pastimes are maintained primarily by elderly Japanese. There is also an impression in Japan that pesticide use is causing a decline in wasps, though wasp harvesters or farmers are unwilling to stop using pesticides for the benefit of wasps.[23] Only a small number of the current generation continue this hobby, despite the work of Japanese *Vespula* societies that seek to promote wasp harvesting and wasp appreciation. Charlotte Payne and Joshua Evans also note that the names of several village-level wasp societies are suffixed 愛好会 (aikōkai), meaning 'loving group'. 'It is clear that many carers have intimate knowledge of, and affection for, their hives.'[24] Only time will tell if these societies are successful.

Pathogens and disease in wasp nests

If the wasps manage to survive the weasels, badgers, buzzards and hungry humans, they still have microbial diseases to keep them awake at night worrying.

Epidemiologists – scientists who study diseases among people – will tell you that diseases are likely to be prevalent whenever there is a high density of highly related individuals who are interacting closely through the sharing of food and body fluids. This nicely describes a wasp nest. Early wasp researchers thought that disease played only a minor role in the dynamics of wasp populations, but signs of disease have long been known to be present in wasp nests. Michael Archer, a long-term wasp researcher in England, observed in 1981 that approximately half of more than 200 common wasp nests had irregularities, including atypical larval development.[25] Within honey bee hives, these irregularities – such as deformed and discoloured larvae, and an oddly distributed brood – often indicate microbial disease. We now know that wasps harbour a diverse community of viruses, fungi and bacteria known to be pathogens in other social insects.

One of our studies here in New Zealand examined viral infections in individual wasp workers from several different nests. We found that every single wasp that we sampled had Kashmir bee virus. This virus was first described in bees and is one of five species that have been linked with honey bee colony collapse.[26] Infection rates varied by a couple of orders of magnitude, but every wasp we sampled was infected to some degree.[27] The variation in infection rate was probably due to sampling wasps of various ages (for instances, older workers might have had more time to build up higher viral loads), or it may even have been that some wasps were super-infected spreaders of the disease – the Typhoid Marys of the wasp world. We've also seen a range of other diseases in wasps, including the fungal disease 'chalkbrood', and the deformed wing virus that has also been linked with colony collapse in honey bees.[28]

Many of these diseases are considered 'honey bee diseases', because they were first examined for and described from these bees. But it's clear that these diseases are found in a range of insects. We've seen Kashmir bee virus as an abundant and prevalent disease in invasive Argentine ants.[29] Scientists have found it in moths, beetles and cockroaches. So it's no wonder viral pathogens are found in wasps too. For all we know, these viruses may have evolved in wasps and spilled over to bees when these insects shared a foraging arena like flowers, or when wasps have raided bee hives for honey. With invasive wasps and ants hosting such a high prevalence of these diseases, it probably means the viruses are more common in the wider environment and in other insects, like our economically important honey bees. This is one of the hidden effects that wasps likely have on many insect communities.

In honey bees, diseases like the deformed wing virus are helped along by parasitic mites like varroa. Varroa mites feed on the haemolymph (the fluid in insects that is equivalent to blood) of late-stage larvae and pupae of honey bees. When these mites feed, they infect juvenile bees with the virus. A bee emerging from its pupal stage may then have deformed wings and be unable to fly. Around the world, varroa and the deformed wing virus cause substantial mortality in honey bees. Wasps, on the other hand, don't suffer from varroa. But there are mites in their nests that appear to be associated with disease transmission. Their nests are host to a range of other organisms, including the mite *Pneumolaelaps niutirani*, discovered by New Zealand researcher Bob Brown in 2012.[30] These little mites, less than 1 millimetre long, aren't known as predators or parasites. Instead, they are thought to feed on the detritus on the floors of wasp nests, and that's where the problems begin. The mites carry *Aspergillus* fungi that may help them digest the rubbish they eat. There are many *Aspergillus* species, with about 60 that can cause

A common wasp queen with the mite *Pneumolaelaps niutirani*. Bob Brown, who discovered the mite in 2011, has found that wasp nests where these mites are present are 50–70% smaller than uninfested nests. We still don't know exactly how the mites affect the wasps, though we suspect they spread disease.

Photo: Bob Brown

disease.* When we examine a nest heavily infested with mites, we see a lot of *Aspergillus* fungi in the mites, as well as several other pathogens. Bob and his team observed the mites feeding around the mouths of larval wasps. They seemed to be eating the dribbles of the little wasps, or perhaps the remnants of the salivary secretions that the juveniles had provided to the workers. Perhaps diseases are transmitted during this feeding. Overwintering wasp queens can sometimes be seen carrying dozens of these tiny mites on their bodies, which effectively becomes the inoculum of mites for next year's nest. So it's clear that these mites have adapted to the life cycle of the wasps, including their overwintering behaviour.

Wasps carry many microbial taxa, including a wide variety of bacteria, viruses, and fungi. While pathogens are present, their microbial communities also include beneficial species. For example, they carry *Lactobacillus* bacteria. You carry species of these bacteria too. But they're beneficial to us – for example, they're a big reason for the growing interest in faecal transplants as a medical treatment for those who have conditions such as Crohn's disease. Poo banks seeking donations from healthy humans are popping up all over the world. So, there is growing knowledge of the positive effects of our symbiotic bacteria and the control they exert over our lives.** But there is much to be learnt about the role of microbial communities in wasps and other social insects. Our work on common wasps clearly shows that quite distinct microbial taxa can exist in larvae, workers and wasp queens from the same nest. How is that possible? Workers share food and most of their biology with the queens – so how do different microbial communities develop? There is a lot that we don't yet know about these wasps.

Earlier I remarked that microbial diseases might keep wasps awake at night, worrying. Well, no, probably not. But you might now be wondering if wasps and other social insects do actually sleep. The answer is yes. And just like people, if

---

* The fungal genus *Aspergillus* was first described in 1729 by the Italian biologist and priest Pier Antonio Micheli. The fungi reminded him of the shape of an aspergillum – a tool used to sprinkle holy water in a church.

** A variety of bacteria are even known to excrete neurotransmitters that act directly on their host's brain. Bacteria thus seem to modulate our response to a range of stimuli, including pain. One study with rats found evidence that gut bacteria could reduce rates of autonomic behaviour, such as gnawing off one's own limbs. I've yet to see many wasps chewing off their own legs.

you deprive them of their sleep they start making mistakes and communicating badly. To examine the effects of sleep deprivation, researchers at Adirondack State Park in New York used an evil-sounding machine called an Insominator on hives of hapless honey bees. They glued little magnetic disks onto the backs of bees. The Insominator was fitted with magnets that would result in the bees with disks being jostled to wake them at various times of the night. The jostled and sleep-deprived bees really suffered in their ability to tell their sisters where they should go for tasty treats – which ordinarily the bees would communicate by doing the waggle dance. This poor communication could thus have dire consequences for food collection and the fitness of bee or wasp colonies.[31] Happily for bees and wasps, there aren't too many Insominators out in the wild.

Successful, mature nests produce new queens and males in autumn

It's now late summer and autumn is approaching. Our wasp colonies have grown, though some nests have done better than others. The nests that were afflicted by disease or overcome by predators or competitors might have only a few dozen or hundred workers. These small nests have a limited ability to produce new queens and reproductive males – perhaps only a handful of new queens, and a few more males. But the nests that have done well can contain more than 20,000 worker and queen cells. These colonies are the 'fittest'. Some very large nests might produce more than 5000 new queens. If the probability of surviving winter is as low as 2.2%, then it's essential to produce a lot of fat new queens to ensure that your genes are represented in the next generation of wasps. Only around a hundred of those 5000 new queens are likely to survive into the next year.

The worst result for any of these colonies is the death of the queen during summer. The death of the queen means the colony can no longer produce new workers, and, even worse, no new queens either. However, not everything is lost. When a queen is alive, she produces a pheromone from her mandibles that inhibits the sexual development and behaviour of the workers. Upon her death, and in the absence of this pheromone, the workers will become reproductively active. But they cannot and have not mated. So the eggs that a worker lays contain only one set of chromosomes. And those eggs turn into male wasps. A colony that has lost its queen can still produce males that might live long enough to reproduce with queens from another nest.

It's the same for the queen, too: if she lays an unfertilised egg, that egg has only one set of chromosomes (directly from the queen) and they are male. However, unlike the workers, a queen gets to decide the sex of her offspring. She has a secret supply of sperm, which she has stored from her mating in the previous autumn. She keeps this in a special organ of her reproductive tract, called the spermatheca. These spermatheca keep the males' sperm alive by providing it with nutrients. Should the queen want to fertilise an egg, a few sperm will be released from her spermatheca. If she lays a fertilised egg, it will turn into a female worker or a new queen. Sperm in insect queens of ant species can be kept alive for decades in these spermatheca. A fleeting, once-in-a-lifetime mating is enough for these queens to produce millions of offspring over many years.

This mating or sex-determination system is called haplodiploidy. Ants, bees and wasps use haplodiploidy extensively, as do thrips (very tiny, slender insects that eat mostly plants), some beetles and true bugs.* Wasp queens and workers have 50 chromosomes and males have 25. In honey bees, the queens and workers have 32 chromosomes and males have 16. (Surely wasps having 50 chromosomes compared to a petty 32 in honey bees makes wasps more interesting?!) A feature of the mating system is that the workers within a haplodiploid nest are highly related to each other. You might have a sister or brother, and though they might appear at times to be from another planet, you share on average 50% of their genes. Sister wasps in a nest, on the other hand, are more closely related to one another because of their haplodiploid system. If a queen has mated only once, the average relatedness between workers in a nest is 75%.

It's this high relatedness that has been suggested as the primary reason that ants, bees and wasps have become 'eusocial', working together in the nest with different tasks and overlapping generations. From the point of view of a wasp,

---

* Most people, with the exception of us ardent entomologists, loosely refer to insects as 'bugs'. 'True bugs', however, belong to the insect order Hemiptera. There are 50,000–80,000 species of true bugs, including aphids and cicadas. All share a common arrangement of mouthparts, which they use for sucking plant or animal fluids.

if you share most of your genes with your sisters, it might be beneficial to you personally to give up your own chance at reproducing in order to increase the overall reproductive success of the nest. This hypothesis for the evolution of eusociality makes logical sense, though is hotly debated and difficult to test. The haplodiploid system has some other neat features. For example, a male wasp has no father. That male also cannot have any sons. But each male has a grandfather wasp. And our male wasp can have grandson grubs.

Male wasps contribute nothing to the maintenance of the colony or the rearing of the young. They have no stingers to help defend the nest from hungry badgers, rodents or humans. The males probably even lead to the decline and destruction of many wasp nests, because when there are a lot of them, their feeding demands are a heavy burden to workers and larvae. Their demands can result in the death of an entire generation of a brood. The late-season brood experiences a double whammy, as wasp workers seem to lose interest in providing larvae with food. The brood failure and destruction of the social wasp colony has been called the *couvain abortif*, which is associated with both the presence of males and the emergence of the young new queens.[32]

The cycle begins again

Male wasps are only good for sex. The adult males will typically leave the nest first, prior to the new queens. They will form mating swarms with dozens or even hundreds of males flying around some landmark or rendezvous place, like a tall tree. Queens might fly into these aggregations. The queens will also release an attractive pheromone that is clearly very appealing and stimulating to males. Even if a queen has recently died, the males will do their futile best with her to pass on their genes to the next generation. The queen's pheromone also appears to rub off onto copulating males, who then, post-coitus, may become of interest to other males. The dense aggregations of males who compete for mating rights with the females is an example of 'sperm competition'. It's best for the female to mate with a strong partner in order to have successful and strong offspring.

Mating is not a lengthy affair – it may take only 30 seconds. In some species, the queen might be captured mid-air and tumbled to the ground with the male.

In common wasps, the male has been described as approaching from behind while on the ground. He fans the queen with his wings and strokes her with his antennae as he climbs on her back. Other males are attracted to the mating scene, and a ball of males can occur with the queen lost entirely from view. Once his half-minute is up, the queen turns on her successful partner, biting him until she is released. The male lives through the experience and might just get lucky again. The queen is also likely to mate again herself. In one of our studies here in New Zealand, we've observed successful queens of common wasps to have mated 2–3 times.[33] Queens in the native range of Europe have a similar number of partners. It's thought that matings of multiple males, or polyandry, are useful because the resulting genetic variation in the nests facilitates resistance to pathogens. The theory goes that some male genotypes are less susceptible to pathogens than others, so having multiple partners is an insurance policy. Because some of your partners might be resistant to disease, some of your children will live if disease infects your nest. Within *Vespula* wasps, there is evidence that the more the queen mates, the larger the nests, and the higher the number of new queens.[34]

Once she has mated to her satisfaction, the queen finds an overwintering site: a place to see the winter through. She has a nice, long sleep. And if she is fat, and very lucky, she might survive to the next year, and the process will start all over again.

2. SPIDERS WITH NO CHANCE OF SURVIVAL

Finding food and feeding

I am in awe of social wasps, both as predators and competitors. They are voracious hunters, and they can learn complex tasks and use novel methods to win their prey. Their learning ability, social lifestyle and high abundance in native ranges must make them one of the most efficient predators in the world.

Decades of research in New Zealand have given us the best data around on the biological and environmental impact of invasive social wasps. In much of this chapter, I'll talk about the wasps in the sub-alpine honeydew beech forests of this quiet but humming little country at the bottom of the world.

Relative survival rates: Old folks vs moths and spiders

I give a lot of lectures about insects. My audiences often contain older people, sometimes people who are 90-plus years of age. To highlight the extent of the predation pressure of wasps in New Zealand's forests, I'll often talk about the chance of someone dying of boredom in one of my lectures. I know it's going to happen one day. In everyday life, the probability of an ordinary 90-year-old surviving to 91 years of age is 85.8%.[1] So, on average, 14.2% of 90-year-olds are going to pass on sometime during their 91st year. The more lectures I give, the higher the probability that someone will snore themselves into oblivion while I'm happily droning on about wasps from behind the lectern. If the chance of our 90-year-old dying remains the same throughout the year, then a crude estimate of the chance of them surviving the hour of my lecture is 1 minus 14.2%, divided by 365 days, divided by 24 hours = 99.8%.

Let's compare this with the probability of moth caterpillars surviving just three hours in autumn in a New Zealand beech forest. For the larger moth caterpillars, that probability has been estimated at just 17%. We know this from a study on a native species called the kōwhai moth, *Uresiphita maorialis*.[2] Kōwhai moth caterpillars are typically green with black and white spots. A 'large' caterpillar might be 15 millimetres long. (By moth standards, 15 millimetres is not very long, as there are many moths around the world that, in caterpillar form, are longer than 100 millimetres.) Kōwhai moths and caterpillars shelter, hide and feed in a range of trees, including the kōwhai tree. Two New Zealand researchers, Jacqueline Beggs and Jo Rees, experimentally placed these caterpillars in trees in a beech forest. Then they waited and watched. Over a 12-hour day, the probability of these caterpillars surviving was just 0.1%. Based on this, they concluded that an estimated 99.9% of large moth caterpillars that are present in beech forests are found and eaten on a daily basis by wasps. Smaller caterpillars fare a little better – when they're present during a time of high wasp abundance, they have a 15%

Beech tree forest in Saint Arnaud, South Island of New Zealand. Up to 40 wasp nests per hectare have been recorded, typically with 5000–10,000 workers per nest.
Photo: Warren Butcher

chance of surviving for a day. But that still means these caterpillars will probably survive only a few days. Growing larger will be to their detriment, serving only to make them more visible to the wasps and to experience an even higher likelihood of being munched. During late summer or autumn, the most optimistic estimate of the probability that a caterpillar would survive the hungry, hyper-abundant jaws of wasps until pupation into an adult moth was 0.0000000000000000000 00000000000000000001%.[3]

In another study, Jo Rees worked with Richard Toft to see how garden orb-web spiders (*Eriophora pustulosa*) fared against wasps. This is a species of spider native to Australia and now present throughout New Zealand. The adults aren't big, as spiders go – they have a body length of about 12 millimetres. The researchers used them in this study because they were easy to collect and would readily build webs on a Y-shaped wooden frame that could be experimentally placed in beech forests. Wasps are known to attack spiders as large as the wasps themselves without seeming to have a second thought. So, sadly, the spiders suffered a similar fate as the kōwhai moths. True to form, the wasps quickly found, attacked, killed and flew off with the spiders.[4]

With larger prey, wasp workers team up. They often attack stick insects and grasshoppers, by first chewing off their heads and limbs. Then the wasps will

A common wasp forager returning to her nest carrying a dismembered spider, as her nest-mates crowd around the plastic entrance tube. This is one of our experimental, observational nests. *Photo: Phil Lester*

chew at the joints between insect or arthropod segments. Larger spiders might have their two body segments – the cephalothorax and abdomen, in spider morphology terms – separated in this fashion. But smaller spiders are killed and then bundled up whole and flown back to the nest.

Richard and Jo found a difference between the survival of small and large spiders. That difference is probably related to the ability of wasps to see or smell the spiders. Larger items of prey are less able to hide, so they are easier for wasps to find. The results of the study indicated that these orb-web spiders had no real chance of surviving a wasp season, which in New Zealand runs from November to June. The most optimistic estimate of a spider hatching in spring and surviving until autumn was 0.000000000000000005%.[5]

When you look at the rates of mortality for spiders and moths compared with that of the older folk who attend my lectures, you can see just how incomprehensibly low the chances of survival are for some of these species.

8 kilograms of insects

Clearly, there are unlikely to be many kōwhai moths and orb-web spiders in the beech forests of New Zealand. Both species are representative of their ilk. To determine the diversity, numbers and weight of insect prey being brought into wasp nests in these forests, researchers tampered with wasp nest entrances to include a tube and container to trap and gas wasps. Readers who don't have a positive view of wasps might now be excited about the idea of gassing them. Unfortunately the gas used isn't some banned chemical weapon; it's just carbon dioxide, which temporarily stuns the wasps.[6]

If you have sufficient patience to watch wasps and intercept their prey over an entire season, you'll find that they are bringing in between 0.8 and 4.8 million loads of insects per hectare. In other terms, that's 1.4–8.1 kilograms of bugs per hectare of forest each year. More than half of these insects are spiders or moth and butterfly larvae. The remaining insects are flies, with winged reproductive ants making an appearance at specific times of the year, and a small assortment of other tasty bugs. Common wasps have a slightly different diet from German wasps. German wasps, which are slightly larger, can carry heavier prey, such as the large native cricket-like species called wētā, found only in New Zealand. (My favourite wētā genus is *Deinacrida*, which translates as 'terrible grasshopper'; it can weigh twice as much as a mouse.) German wasps in these forests were the only species observed killing and carrying honey bees back to their nests.[7]

That 1.4–8.1 kilograms of insect prey consumed by wasps has been estimated as similar in both composition and weight to the entire prey intake of all insectivorous birds that are normally found in these areas. These birds include the small riflemen, fantails, warblers, whiteheads, endangered yellowheads and other species.

Also on the wasp menu are vertebrates – animals with backbones. Wasps have long been known as scavengers, foraging on a wide variety of dead animals. A gruesome list of items by the field naturalists Phil and Nellie Rau, writing in the 1910s, included the eyes of dead rats and the decapitated heads of roosters. One rooster's head was described as 'teeming with these greedy savages. They jostled and swarmed over the tongue, eyes and all exposed moist surfaces'.[8] It is probably much rarer for wasps to attack live vertebrates than to scavenge or hunt invertebrates, but large, mobile protein sources are hard for them to ignore. For wasps, birds are tasty prey, especially nestlings. A correspondent to an early issue of *British Birds* describes wasps forcing adult birds from their nests in order to

Wasps foraging on a rabbit carcass. Wasps probably play some role in decomposition and the recycling of nutrients, but without wasps other species would take on such a role. *Photo: Dave Hansford*

gain access to chicks.[9] All the parent birds could do was sit on a nearby branch, watching and 'churring' as the wasps stung and killed their offspring before they 'gnawed at the young nestlings to obtain loads of meat'.[10] Such killings have also been seen here in New Zealand, where newly hatched nestlings of introduced sparrows and native robins have been devoured by wasps.[11] The adult birds make alarm calls and swoop down on their nests that are under attack, but to no avail. Eventually there are just little chick bones in the nest. Some observers have suggested that the smell of the egg albumen might attract wasp foragers.[12] The occasional adult bird can become dinner for a wasp as well – wasps in the related genus *Dolichovespula* have been observed attacking and killing adult hummingbirds in mid-air.[13]

German wasps are also known to prefer even large live prey, chewing on the teats of dairy cows. Other wasps have been seen attacking and eating tadpoles. The most extreme animal consumption, with perhaps a little ironic revenge, must be that of wasps eating an entomologist. A dedicated entomologist in Canada once allowed a *Vespula* forager to gnaw on his ear until it drew blood, after which the wasp swallowed the blood and flew off.[14] That's either dedication, or something else.

The studies on kōwhai moths and orb-web spiders both found that predation by wasps is density-dependent. That is, the more wasps you have, the higher the probability that a prey item will be discovered and quickly killed. These results highlight the major issue with wasps in these forests. The issue isn't that these wasps are present. There are many exotic species in these forests that don't seem to be an ecological problem, simply because they are rare or in low densities. If there were only one wasp nest every 200 hectares, wasps wouldn't be as big a threat to spiders and moths. The issue for wasps is that there are just so many of them. With up to 40 nests per hectare, and each nest housing thousands of hungry and demanding mouths to feed, wasps are a ferocious feeding force. Richard and Jo estimated that wasp numbers would need to decline by 75–90% in order to increase the survival rate of the Australian orb-web spider to just 0.1%.[15] This survival rate would enable the spider to persist in these forests, because it has a high reproduction rate. New Zealand native spiders might need an even steeper decline in wasp numbers, because they have a lower rate of reproduction. Similarly, the study by Jacqueline Beggs and Jo Rees estimated that the kōwhai moth would need wasp densities to go down by somewhere between 85 and 91% before it could survive times of high wasp abundance.[16]

With survival probabilities of 0.000000000000000005%, you'd expect to see extinctions. Insect populations, however, are generally not subject to repeated and extensive surveys. People usually don't spend decades gathering the necessary data to show extinctions – they would need to survey moth or spider populations over thousands of hectares over many years. And even with such surveys, extinctions are hard to prove. Large animals previously pronounced extinct have later been rediscovered. Pygmy whales, musk deer, kunimasu salmon in Japan, and mountain dogs were all at one point or another thought to have been wiped from the face of the earth, only to be rediscovered alive and well. (These animals are sometimes referred to as Lazarus taxa.)

Within New Zealand, we have only anecdotal evidence that wasps have at least greatly reduced the abundance of some native insects.[17] A survey by the New Zealand Forest Service sampled moths in a forest for five years prior to a common wasp invasion, then for six years afterwards. Species such as the endemic forest looper moth (*Pseudocoremia suavis*), which has an adult wingspan of about 30 millimetres, were abundant before the wasps arrived. As a caterpillar, the forest looper is large, juicy and abundant during late summer. Before wasps, the looper would occasionally reach epidemic or plague numbers, eating its way through the foliage of a vast number of trees. But after the arrival of wasps, the Forest Service were able to find only a few loopers over the entire six-year sampling period. Those in the forestry industry might argue that the dramatically reduced abundance of moths and their tree munching is a good thing.

Not all insects fall victim to tiny wasp jaws. Species that breed in the spring or outside late summer and autumn, which is the usual period for high wasp densities, will likely escape. Few wasps are around to bother chicks that hatch in the spring or early summer. Similarly, a beetle that breeds and lives in rotting logs or under thick layers of leaf litter won't often be discovered by wasps. And because wasps sleep at night, they miss out on the nocturnal creatures that hide during the day and forage at night. So we cannot generalise the effects of wasps on all insects and birds. The most useful generalisation we can make is that the species most likely to be attacked by wasps are those that are most often around and having their young in the late summer or autumn.

An abundance of sugar fuels the wasp fire

These insects and the other prey are a protein source that wasps need for feeding their larvae. In New Zealand beech forests, the energy source for wasps is native scale insects (*Ultracoelastoma* species), which are present in massive densities. These native forests, each covering about 1 million hectares, can be home to huge densities of scale insects. Individual trees can have up to 1535 scale insects per square metre of tree bark.[18] Another estimate suggests that there are about 20 million naturally occurring scale insects per hectare of beech forest.[19] Scale insects suck phloem sap from the tree. The sap is high in sugars but low in nutrients such as nitrogen and amino acids, so the insects need to process a lot of it to sustain themselves.

Because of all the sap they're eating, the insects produce a lot of sweet, sugary honeydew as a waste product. When unharvested, the honeydew falls to the forest floor or against the tree trunks. There, the sugary exudate cultures a sooty mould that stains the trees black. These forests look a little like a crowd of glittering Christmas trees, because the anal tubes of the scale insects extend 20–30 millimetres perpendicular to the trees. Each tube holds a little droplet of honeydew that glistens in the sun. The amount of honeydew produced is staggering. One estimate put the honeydew production rate at between 3800 and 4600 kilograms per hectare, per year.[20] In the absence of wasps, you can walk through these forests and the air literally smells sweet. You can even touch the anal tubes of these scale insects to gather a little honeydew to eat. It's very sweet.

If consuming the anal excretion of these insects sounds gross, just think about honey. 'Bush honey' in New Zealand comes from bee hives that are positioned in native forests. Honey bees, if given the chance, will readily collect the honeydew from scale insects then take it back to their hives to mix with nectar from flowers. This mixture is refined into honey by individual bees repeatedly drinking, then vomiting, the increasingly concentrated sugary substance into the comb cells. Honey around the world is frequently a violently expelled end product – from both ends – from a variety of insects. And people have been eating these end products for thousands of years. A well-known example is the 'manna from heaven' that the Israelites lived on following their hasty exit from Egypt. A prominent theory is that this manna is the dried, crystallised sugary excretions from Tamarisk manna scale (*Trabutina mannipara*), still considered a delicacy in parts of the Middle East. So this insect waste product may have sustained an entire nation! The Israelites were under strict instruction to eat the manna soon after collection and certainly not leave it until the morning after. Stored manna would breed worms and rot, and make Moses wroth.[21] If the theory is correct, it's likely that the manna had eggs of these 'worms' when it was collected. The eggs might even have contributed to the nutritional value of the manna – just as the few small insect eggs or body parts you likely eat in your muesli every morning contributes a little to your daily nutrition.*

Honeydew in New Zealand beech forests is an abundant, high-energy product. It fuels native honeyeater birds such as tūī, bellbirds and silvereyes. And unfortunately it fuels wasp populations as well, and wasps dramatically alter the quality and quantity of the honeydew. Experiments have shown that the native honeyeater birds have little effect on honeydew quality and quantity. But wasps are another story. In the summer and autumn months, when there are plenty of wasps about, the number, size and sugar concentration of the honeydew droplets is massively reduced. One group of researchers found that the energy in drops per unit area of tree trunk was reduced by 91–99% from January until May. They even suggested that these measurements were an underestimate of the effects of wasps on honeydew, because of limitations in their experimental design and ability to estimate honeydew consumption. One tree in their study was seen with 370 wasps per square metre of its bark. The wasps were so good at harvesting the honeydew that the authors of the study had a hard time finding enough honeydew to enable them to take measurements. 'Wasps reduced the standing crop (number, size, and sugar

---

* It's virtually impossible to keep insects out of your diet. Take figs – they exist because of fig wasps. The approximately 1.5-millimetre-long female fig wasps emerge in the mature figs of one tree, become coated in pollen, then fly to another tree. A female penetrates the immature 'fruit' (technically they are inverted flowers) to lay her eggs, and inadvertently she pollinates the plant. Her eggs hatch and the baby wasps eat a small amount of fig. So, when you eat a fig, frequently you are eating some little wasps or wasp remnants. Because of this, figs are off the menu for strict vegans. Organisations such as the USFDA place limits on the amounts of insects in foods. Fig paste must have no more than 13 insect heads per 100 grams. Ten wasp heads are just fine.

A common wasp foraging for honeydew on a beech tree (*Fuscospora* species). You can see the anal filaments of three scale insects (one has a droplet of honeydew on its end). The bark is black from the sooty mould that grows on unharvested honeydew.

Photo: Phil Lester

concentration) of the honeydew drops by cropping them before they could be fully recharged and before evaporation could concentrate the sugars,' they observed, with frustration.[22]

Invasive wasps compete efficiently for both sweet honeydew and insects in these forests. If the wasps are harvesting honeydew so efficiently, and eating approximately the same weight of insects as the entire bird fauna, it seems a logical conclusion that they are limiting the food that birds can eat in these areas.

But is there evidence that wasp competition affects native birds? One bird species for which we have good data is the endemic forest parrot, or kākā (*Nestor meridionalis*). Usually, you'll hear kākā before you see them. Their screeches and whistles echo around a forest for miles. Kākā were once widespread in New Zealand; trees could teem with them, and they were an important source of food for Māori. Early European settlers hunted them as well. Today, kākā are among New Zealand's endangered birds, primarily because introduced mammalian predators, such as rats and possums, like to eat them. As an omnivorous parrot, the kākā eats honeydew, and it also has a longing for large caterpillars that hide

Left: A honeydew beech tree when wasps are absent. Each of the hair-like strands is an anal tube of a scale insect, typically with a droplet of honeydew on its end. The tree is black from the sooty mould that grows on the bark from uncollected honeydew. Right: Common wasps foraging for honeydew. They flock to the sugary exudate and exclude native birds. Birdsong is absent from these forests in autumn; it's replaced by the drone of wasps.

Photos: Julien Grangier

inside trees. The kākā tears trees apart and can even kill them as it forages. Honeydew is an important source of energy for this bird, because foraging for caterpillar protein uses up more energy than it gains. The kākā can obtain all of its daily energy needs from honeydew within less than four hours of foraging, at least when wasps aren't around.

Wasps have been suggested and examined as a major limiting factor for kākā abundance, due to their competition for honeydew. One study found that for about four months of the year, the amount of energy contained in a drop of honeydew dropped so dramatically due to wasps that, for the kākā, it was no longer worth foraging for. The sheer abundance of wasps means that they efficiently harvest the honeydew before kākā can even find it. Wasps also directly interfere with kākā foraging, as they are particularly aggressively foragers.[23] The authors of that study suggested that the loss of honeydew to wasps was a big reason why they saw only one successful kākā nest in their six-year study. Introduced predators including stoats, weasels, possums and rats were another important reason, and a main cause of bird mortality.

What is the effect of wasps on other birds? During a long-term study on bird abundances in honeydew beech forests, common wasps arrived into New Zealand. Led by Graeme Elliott, the study was examining the abundance of 11 bird species between 1974 and 2007. Graeme's group found that following the arrival of the wasps, five bird species declined in abundance, three rose, and another three showed no statistically relevant change in density. The five whose numbers dropped were all birds that either only ate insects or relied heavily on invertebrate prey during their breeding season. These birds were more affected at lower elevations, where wasps were more abundant. Graeme's study indicated that no bird species were totally excluded or went locally extinct from these forests. But birds like the rifleman, New Zealand's smallest bird, did suffer, declining to densities less than a third of what they were at the beginning of the study. From the survey data, it is impossible to conclusively link these declines in bird populations specifically to wasps. For example, there are other introduced animals in these forests, like the Australian brushtail possum, that have been caught red-handed (literally) eating chicks. However, even with plenty of other hungry predators, it seems very likely that the aggressively competitive wasps have contributed to the declines of an array of birds in these forests.[24]

Bryce Buckland, who helped to found a conservation group named Friends of Rotoiti, lives near where this long-term study was conducted. What Bryce offers is less scientific and less statistical. He offers instead the experience of someone who has lived near wasps for many decades and who has clear observations on wasp–bird competition. 'Years ago, before we had wasps, this area was just bursting with birdsong,' he explained in an interview with his local newspaper the *Nelson Mail*. 'All the birds had plenty to eat, but wasps took care of that. The birds get really hungry.' Bryce provides sugar-water feeders for the birds during the summer. 'I fill up that 3-litre container probably three times a day. Often the tui and bellbirds are so hungry they just fight like cats, they're just in there bloody scrumming for it. They'll empty that 3-litre container in 15 or 20 minutes.' The recent deployment of an effective wasp control programme using the poisoned wasp-bait Vespex in the forests near his property has made a big difference. Bryce has seen many more birds around, especially the honeyeaters. He believes this is because of reduced competition with wasps. 'We may have 50 or more – it's hard to count them – tui and bellbird feeding on sugar water and within a day or two after Vespex is applied we have none of these at all. This suggests that their usual food supply is then available and a "handout" by way of sugar water is no longer needed.'[25] When their wasp competitors are gone, the birds can find enough sugar on their own.

Wasps alter underground microbial communities

An annual honeydew production of between 3800 and 4600 kilograms per hectare per year[26] is a lot of sugar. In the absence of wasps, some of that sugar is consumed by birds. Without wasps, the vast majority of the sugar, however, is washed by rain or blown by wind to the forest floor. This honeydew can clearly influence microbial communities, which you and I can see only because of the sooty mould. Sooty mould is a collective term for various fungal species. It is estimated that at least seven species of this fungi will grow over any surface where the sweet honeydew falls. The forest is blackened by the mould. I don't think that the forest could look any blacker or burnt, even after a forest fire. Sooty mould is the most obvious effect of the honeydew addition, which really represents the limit of the casual observer's ability to understand the effects of this sugar addition on the microbial community. It took a four-year experiment and a heck of a lot of detailed analysis to get a real handle on the massive impact that wasps and honeydew can have on this system.

Ecologist David Wardle and his research team are attempting to understand how above-ground and below-ground communities interact, and their work offers our best understanding of how wasps can influence microbial communities. So what are the consequences of wasps above-ground on below-ground communities and the ecosystem? The team carried out a four-year experiment to determine how the addition and removal of honeydew could change the decomposer communities. The removal of the honeydew simulated its harvesting by the invasive wasps. They examined the influence of this sugar on key nutrients such as carbon, phosphorus and nitrogen, as well as the abundance and community composition of bacteria, fungi, nematodes and tardigrades (also known as water bears).[27] Most people will have heard of nematodes, or little roundworms, generally a fraction of a millimetre in size. Some of you won't know water bears. Possibly of some disappointment to many, water bears aren't polar bears in a bath. Adult water bears are about half a millimetre long and may be plant eaters, bacterial specialists, predatory or even cannibalistic. (They are cute, though.) And they can survive temperatures of near absolute zero (-273°C) and up to 150°C. They can survive being shot into space with no oxygen. They can go without food and water for 30 years and still live. These tiny bears have much to teach us.

The addition of sweet sugary honeydew clearly shapes the microbial community and its consumers in honeydew beech forests. It promotes a lot of fungi at the expense of the bacteria. When you take away the honeydew, you end up with community in which bacteria play a much higher role. And as you'd expect, a forest that has a lot of fungi attracts fungus-eating organisms. In places with lots of mushrooms, you'd expect to find animals that eat mushrooms. So in these fungi-rich beech forests we see a high diversity and abundance of fungal-feeding nematodes. The authors draw an analogy with changing a TV channel: the addition of honeydew causes 'a switch from the "bacterial-based" energy channel to the "fungal-based" channel'.[28] The addition of sugar caused some organisms to increase and others to decline. Honeydew increased the abundance of predatory nematodes (which feed on microbe-feeding nematodes) and water bears. But it also seemed to reduce numbers of mites, fly larvae, dipteran larvae, and the overall diversity of the arthropod community.

The results of this study show multi-trophic effects that demonstrate that what happens above ground can strongly influence what goes on below ground. One of the most striking findings of this study was the effect on the storage of carbon, nitrogen and phosphorus in the humus of the soil. David and his team predicted that the invasion of wasps may lead to an increase in humus carbon sequestration by around 38% over four years.[29] Wasps drive this increase in carbon acquisition by changing the microbial communities. In a world where we are – or should be – desperately trying to limit the amount of carbon in the atmosphere, you could consider this carbon sequestering to be a positive effect of wasp invasion. David and his team found that high densities of wasps results in the microbial community in the soil sequestering an additional 0.4 kilograms of carbon per square metre of ground, compared to when there are no wasps.

After reading these estimates, I found my calculator. Given that there is a million hectares of honeydew beech forest and high wasp densities in New Zealand, this would scale up to 4 million tonnes of carbon sequestration – simply due to the presence of wasps. Given that New Zealand's total CO_2 emissions are approximately 80 million tonnes per year, that's a significant amount. I showed David my calculations and figures. He pointed to very similar results from his research on rats invading offshore islands. He and his team have found that, by preying upon seabirds, rats have indirectly enhanced carbon sequestration. Live plant biomass increased by 104% and total ecosystem carbon storage increased by 37% due, indirectly, to rat predation of birds.[30]

'I think this highlights the problems of using ecosystem services as a major justification for conservation,' David told me. 'If maximising carbon sequestration was a major goal for managing our natural ecosystems, then these results would provide an incentive to encourage the spread of both invasive wasps and invasive rats, despite their obvious and severe threat to native biodiversity.' David also raised a note of caution around these extrapolations. The wasp study was done in an area of many, many wasps, and the results are from just one location among the million hectares of honeydew beech forest. Results could differ in other areas. My back-of-the-envelope calculation of 4 million tonnes of carbon sequestration due to wasps in beech forest is probably an overestimate too.

Small brains, but really smart

Hopefully by now you'll be convinced that invasive wasps are competing with native animals for resources. I think the evidence is clear that their foraging can substantially change the environment. But you might not be so convinced, yet, that a major reason for their success is the immense intelligence they've shown in going about their work. Wasps have tiny brains: most social wasps that have been examined entertain a total brain volume of under 0.5 cubic millimetres.[31] Some species have as little as 0.08 cubic millimetres. Whatever its size, a wasp's brain has distinct regions that perform different functions. Some regions deal specifically with chemical input from the antenna. Others deal with visual information. Wasps and other insects also have highly developed brain regions called mushroom bodies, which play a major role in cognitive processing tasks, such as spatial orientation and navigation, sensory integration, learning and memory.[32] They might have little brains, but they are very smart.

When hunting, wasps use their full range of senses. Their ability to learn to home in on environmental cues was highlighted in a study of German wasps foraging on Mediterranean fruit flies in Greece.[33] When attempting to entice females to mate, male fruit flies hang out together in groups called 'leks'. In a lek, the males push one another around, act tough and generally dominate their street corner of the tree. They also release a pheromone to attract females, and display visual and acoustic signals when the lady in question arrives. Lek sites are often in the middle of trees or dense foliage. German wasps have learnt to follow the pheromone trails of male fruit flies. They will fly upwind, zigzagging through the foliage and the pheromone trail until they find the lek site. When they're close, they use their eyesight and the sounds of the male flies to locate their prey – then they pounce. No sting required. Just grab, wrestle and bite.

If the food is too large for a single trip, wasps use spatial cues to find their way back and forth to the nest. We now know that wasps can memorise distinctive landmarks including large stones or posts, and use orientation flights to familiarise themselves with specific environments.[34] In a scenario where a wasp has just killed some prey that is too big to carry back to the nest in one trip, an orientation flight will begin while the wasp is leaving the scene. The wasp faces its prey and flies back and forth in arcs of an increasing radius, gradually increasing its vertical and horizontal distance from the prey. It completes its departure flight by circling high above the ground. Only then will it fly back to the nest. The wasp seems to be looking for objects to align with and for images to remember and match when returning to the site. With each return trip to the prey the flight path becomes simpler, until it becomes almost straight.[35]

The ability of wasps to memorise landscapes and the location of food is impressive. We see a wide range of navigational skills and abilities in the insect world as a whole. One of the most interesting navigational methods is that of the dung beetle. Dung beetles forage for dung at night, en masse. With so many beetles desiring a dung deposit, there is substantial competition and a need for a fast and efficient retreat. 'The dung beetles don't care which direction they're going in,' writes one dung beetle researcher, 'they just need to get away from the bun fight at the poo pile.'[36] The beetles are known to roll their collected balls of dung in straight lines away from the bun fight. How? Only recently have they been shown to use stars in the Milky Way. The beetles seem to use differences in brightness and colour as a compass cue to hold a straight line.[37]

Wasps are not yet known to do similar things. Many hornet species fly about on bright moonlit nights, though it's their large size and big eyes that seem to help them find their way.[38] Honey bees are a classic example of insects using the direction of polarised light (linear travelling light waves from the sun, which we can't see) as a major cue for navigation. Wasps and ants have similar eyes, and it's likely that they use polarised light extensively too.

A recent study of hymenopteran learning featured bumble bees. In 2017, researchers from Queen Mary University of London decided to determine if bumble bees were smart enough to learn new tasks well outside of their natural behaviour.[39] The study's co-author, Clint Perry, explained, 'We wanted to explore the cognitive limits of bumble bees by testing whether they could use a non-natural object in a task likely never encountered before by any individual in the evolutionary history of bees.' So the researchers had the bumble bees play soccer. In a series of experiments, the bees would receive a sweet, sugary reward if they rolled a little yellow ball into a goal. Some watched as a trained bee completed the task, some received a 'ghost' lesson where a hidden magnet was used to move the ball around, and some bees received no lesson at all – they just found the ball already in the goal, with a reward waiting for them. Most impressive were the bees that were observers only: these bees were not directly trained to roll the ball to a goal for a reward. They were not only able to learn from the trained bees, but also to enhance their own efficiency and technique. 'The bees solved the task in a different way than what was demonstrated, suggesting that observer bees did not simply copy what they saw, but improved on it,' said co-author Olli J. Loukola. 'This shows an impressive amount of cognitive flexibility, especially for an insect.'[40] The bees' behaviour showed evidence of independent thinking. They would move the ball closest to the reward rather than one a demonstrator was moving. They would roll different coloured balls and not just the ones they had been trained on. The authors concluded that the bumble bees' cognitive flexibility – their ability to learn new tasks, and improve on ways of doing things – hints that 'entirely novel behaviours could emerge relatively swiftly in species whose lifestyle demands advanced learning abilities, should relevant ecological pressures arise.'[41]

With invasive wasps, we've seen what we believe to be just this sort of novel behaviour under new ecological pressures. In New Zealand forests, the native bush ant (*Prolasius advenus*) is highly abundant and dominant on the forest floor. If a dead insect falls to the ground, these bush ants will usually be the ones to eat it. At least, that is, in the absence of wasps. The fun begins when ants and wasps are competing for the fantastic prize of an insect carcass. The ants are tiny: a single wasp is more than 200 times heavier than one of them. Nevertheless, the ants are aggressive. They will rush at a wasp and bite its ankles or spray acid in its face. What should a wasp do? In order to clear the deck of ant competitors, the wasps very quickly and carefully grab and airlift the ants away from their desired lunch. In a fraction of a second, a wasp will dart in, pick up an ant with its mandibles, fly backwards, and drop the ant away from the food. The ant appears a little confused – 'What just happened?' – but seems otherwise undamaged from the experience. The wasp does its airlifting gently, probably because each ant really represents a loaded chemical weapons factory. Each ant is armed with formic acid. Should the wasp bite too hard, it will have a nasty taste in its mouth for the next several hours. No ants are eaten: they are just removed. Impressively, the more ants that are teeming over the food, the greater the distance the wasps fly them away. The lead author of a study on this behaviour, Julien Grangier, notes that, to the best of his knowledge, it hadn't been seen before. 'Our results suggest that these insects can assess the degree and type of competition they are facing and adapt their behaviour accordingly. It's a new interaction between a native and an invasive species and a wonderful example of behavioural plasticity.'[42]

It would be wrong to leave you with the impression that all wasps are fantastically smart hunting machines. Wasp smarts show a bell-curve like all other animals. There are some super-smart behaviours in the wasp world, but we also come across behaviours that would suggest . . . well, a lesser intellect. While some clever wasps will locate their lunch by using the sex pheromones of their prey, there are orchids that produce wasp sex pheromones to convince male wasps to repeatedly pounce on and rut at their flowers. The flowers look nothing like female wasps, and the males get no reward for their misplaced reproductive efforts. All that happens is the flowers get pollinated. Wasps have also been known to mistake nail heads for prey – and will repeatedly pounce on the barbs of barbed wire fences. Why? There is one theory that wasps have a habit of attacking objects with irregular profiles.[43] They will also attack, beat up and rob their nest mates of prey, even though that same piece of food was going back to the same nest to feed their same sisters. Some morally bankrupt wasps

Wasps and native ants interacting near an experimentally placed food item (we used tuna fish). If there were just a few ants, the wasps would share the food, ignoring any aggressive behaviours from the ants. If there were many ants, the wasps would pick up each ant and drop it elsewhere. The ants appeared unharmed by this. We think the wasps were probably careful because if they bit too hard, they'd get a mouthful of acid. *Photos: Julian Grangier*

seem to specialise in beating up and stealing from their siblings, rather than finding their own prey to take back to the nest.[44]

The 'murderous hoards'

A lot of the studies I've talked about here are about wasp populations that exist in extremely high densities. And so many of these results regarding the effects of wasps on biodiversity are among the most dramatic or are even the worst-case scenarios. We certainly don't see densities of 370 wasps per square metre of tree trunk everywhere throughout their invaded range. And without such high densities, wasps could not take 99% of the available honeydew or consume the equivalent biomass of insects as all of the resident native birds. Where they are present in relatively low densities, they don't cause such big problems. But these invasive wasps are extremely abundant over large areas of their invaded range, including over the 1 million hectares of honeydew beech forest in New Zealand. They are major competitors and predators in Hawai'i, Chile and Argentina. The major biodiversity and conservation problems arise when we have masses of wasps.

If you're a fan of wildlife documentaries, *Bandits of the Beech Forest* will take you into the high drama of the New Zealand wasp invasion. It will give you a visual and auditory experience of the buzz of these wasps in New Zealand's beech forests.[45] *Bandits* won the Environment Prize for Best Film Illustrating Protection, Preservation or Conservation of Bird Life at the Festival du Film de l'Oiseau in 1997, presumably for such scenes as a large stick insect being torn apart by wasps, starting with its head being chewed off and separated from its body, and wasps gulping the honeydew from the scale insects on blackened trees. 'These are the most efficient predators the New Zealand forests have ever seen,' intones the narrator. 'Working in murderous hoards, they dismember victims far larger than themselves. The songs of the kākā seem destined to be silenced forever. For those who have heard it, it is of great sadness to imagine the beech forest without their song. . . . Over 50 years, the bandit queen and her kind have now imposed a new song in this forest.'

3. 'I PUT MY FOOT RIGHT IN THE NEST!'

The costs of wasps

Janet Kelland was walking along a fence line on her remote sheep farm when the ground collapsed beneath her feet. She had stepped into a wasp nest. The nest was underground, near the base of the fern. Over the next few minutes she would receive hundreds and hundreds of stings. Each and every wasp was using its sharp stinger to repeatedly pump venom into her body. The 56-year-old thought she was going to die.[1]

Understandably, Janet wasn't able to look closely at the wasps to identify the exact species. We don't know if they were common wasps or German wasps. To Janet, they were just wasps: yellow and black, about 2 centimetres long, and angry. Common and German wasps both like New Zealand very much. They have flourished in this temperate habitat that once was devoid of any native social wasps or social bees. The two species of social wasps have now spread throughout the country and often produce very large nests with tens or even hundreds of thousands of workers. And these wasps really resent uninvited guests, such as Janet's foot, entering their homes and squashing the kids.

'I was just covered in wasps,' Janet later said. 'They were stinging me through my shorts and shirt.' She had been marked as a villain. The first wasp that stung her released not only venom but also a pheromone that alerted other wasps that she was a target for stings and extreme aggression. These additional stings served to increase that marking and aggression. 'They were in my bra, in my underpants. I was pulling handfuls of wasps from my hair. . . . It was just incredible, they went on and on, and on, and on. There was layer after layer of wasps.' Janet tried to get rid of the wasps by running into the forest and

66

rolling on the ground. 'I tried to squash the little buggers.' This just made them angrier. She ran down a slope and finally escaped their fury by diving into a small stream.

The attack was over. But after hundreds of stings, Janet was by no means out of trouble. She was alone on the farm, and the effects of the venom were just kicking in. Her first task was to walk back up the hill towards the nest to retrieve her motorbike. She would then need to ride to an area with cellphone coverage and call for help. All up, this was a trip of nearly one hour. 'I knew I was in the shit, but my throat didn't swell, so I could breathe.' She was determined not to die. 'If I died there no one would find me. And the wasps would eat me. Then the flies would come. And then the pigs.' (Feral pigs, another introduced species to New Zealand, are a major scavenger and pest in many areas, and are also on the list of the world's worst one hundred invasive species.)

Finally, she reached an area where she could use her cellphone. Her next task was to convince the emergency operator that her location and farm (in Taumarunui, in the central North Island) actually existed, as the operator couldn't find her on a map. That must have been an interesting conversation. Eventually a rescue helicopter arrived with medication and cold wet towels to calm the pain. Did it hurt? 'It was painful at the time, yes, but mostly later,' said this master of understatement. After so many stings, her doctors said that she should have died. A person who was not as fit and as determined as Janet would likely not have survived. Six months later, her back was covered in scars. 'I was lucky – very, very lucky. I shouldn't have survived that ordeal.'

Mortality rates

Janet's story had a happy-ish ending. Of course, she would prefer to have side-stepped the nest and not be stung at all, but she did live to tell me the tale.

Other people haven't been so lucky. The first recorded human fatality from a wasp sting is frequently quoted to have been the Egyptian pharaoh Menes, around 2600 BCE. We will never know the truth for sure – it is possible, for example, that the poor pharaoh was sat on by a hippopotamus, or that his own dogs chased him into a lake to drown, or that he was eaten by a crocodile.[2] But for the sake of a good story, let's assume he was stung by a wasp or a hornet and consequently died from anaphylactic shock.

More recently and closer to home, in 2012 a 62-year-old man named Morris Stretch was collecting firewood with his nephew in Kenepuru Sound, the upper South Island of New Zealand, when the pair accidentally disturbed a huge subterranean wasp nest.[3] He and his nephew ran in different directions to escape the wasps. But Morris was stung repeatedly. When his nephew came back to find him and opened his shirt to give CPR, Morris still had wasps in his shirt and all over his neck and face. He died of a heart attack. Morris probably received hundreds of stings.[4] But if you are extremely allergic to wasps, just one sting can kill.

A study from 1994 estimated that the annual rate of mortality from hymenopteran stings was somewhere between 0.09–0.45 deaths per million people.[5] Note that this sort of rate was for hymenopteran stings, which includes stings from honey bees, hornets and perhaps even a few species of ants, as well as the solitary and social wasps. This lumping together of hymenopteran 'villains' is an unfortunate outcome of the public – and even medical doctors – not being able to distinguish bees from wasps. An entomologist such as myself sheds a tear or two upon hearing of this confusion, though I do appreciate that, at the time, non-entomologist muggles are more interested in stemming the pain than in finding out its origin. The mortality figure from 1994, however, was almost certainly an underestimate. Hospitals will often record a cause of death as a cardiac arrest even when a heart attack has been brought on by an allergic reaction to a sting. Morris was just such an example: his cause of death was classified as coronary failure, but in the absence of wasps he likely would not have had a heart attack at that particular time. After accounting for the confusion over cardiac arrests as an indirect result of stings, or the more direct lethal effects of anaphylactic shock, the most recent estimate of the mortality rate from common and German wasps in New Zealand is approximately two deaths per year, for a population of 4.4 million people.[6]

Two deaths a year in a population of 4.4 million doesn't sound like much. Right? Well, consider Australia. There are some nasty animals in Australia: the most venomous snakes in the world, spiders lurking under the toilet seat just itching to bite your bare bottom, and tiny jellyfish with miniature explosive harpoons of poison that will stop your heart in seconds. In 2017, an Australian research group put the mortality rate from hymenopteran stings into the context of all venomous injuries that include these spiders and jellyfish. They found that, despite all these venomous toilet-lurking creatures and sea monsters with exploding harpoons, the highest cause of hospitalisation and the first-equal cause of mortality in Australia was hymenopteran stings. From 2000 to 2013, stings from ants, bees and wasps killed more Australians than sharks and crocodiles combined. The study found that, typically, the cause of death is not the stings themselves, but allergic reactions.[7]

These allergic reactions can only be described as an overwhelming, unnecessary and illogical response from our immune systems. If you wanted to define and exemplify 'overreaction', a person's anaphylactic response to a few microlitres of wasp or bee venom surely takes the cake. The amount of venom a wasp injects is tiny. And yet, as we'll see later, this impossibly small amount of toxin can initiate a dramatic and even fatal reaction in people and animals.

Breaking down the Australian mortality statistics even further, you will see that the mortality rate is almost twice as high for men as it is for women. For some reason, men appear to be more prone to wasp stings and attacks. In fact, New Zealand men have been estimated to be three times more likely as women to be on the sharp end of a wasp sting.[8] Behavioural differences are, well, probably responsible. In a 1993 study about the differences in the behaviour of men and women towards wasps, one group of subjects nicknamed the data set the 'wasp sting stupidity chart'. Due to exhibiting less wariness around wasps and wasp nests, young men typically rated highly on this chart. As the authors of the study concluded, 'It may well be prudent for young males to exercise caution when dealing with wasps.' The words 'prudent' and 'young males' aren't often seen in the same sentence together – but, in defence of young males, risk-taking is apparently in their nature. A wide variety of studies have concluded that our male ancestors needed to take risks in their hunter-gatherer roles in order to put food on the table; whereas females, theoretically, developed more prudent and cautious traits as they worked to keep families protected and together. So apparently the high rating that men receive on the wasp sting stupidity chart is a function of their genes more than a conscious behaviour. Genes are an oft-used excuse for questionable male behaviours, but this one stings.

It's worth noting that other male mammals have similar intellectual challenges that led them into trouble. Love Dalén, an evolutionary biologist from the Swedish Museum of Natural History, recently led a study on historic mammoth deaths. He found that male mammoths were much more likely than female mammoths to be, for example, swallowed by sinkholes or washed away in mudflows.[9] 'In many species, males tend to do somewhat stupid things that end up getting them killed in silly ways,' Dalén said, 'and it appears that may have been true for mammoths also.'[10] It wouldn't surprise me if male mammoths received the lion's share of wasp stings for their species as well.

Invasive species can hurt your pocket

We know that invasive wasps are a critical issue for New Zealand's biodiversity, especially its insect communities. The impact of wasps on our biodiversity rates alongside that of rats, and of the diseases and pests of honey bees.[11] Most people don't see these effects, though. If you don't often travel to native forests, don't spend much time outdoors, or don't have bee hives, you probably won't have many interactions with social wasps. Wasps will generally leave you in peace if you're sitting in front of a screen indoors. For most city-dwelling people, wasps become relevant only when little Susie is stung at the local playground and has an allergic reaction. Then your family gets interested, and possibly even the media, often fuelling insect phobias and fears.

Biological invaders can seriously affect the economies of their host countries. We are often told as much by the news media, with stories such as 'Invasive non-native species cost the Scottish economy at least £250 million each year', and 'Great Lakes invasive species cost US$200M a year'.[12] Accurately estimating the cost of invasive species, however, is very difficult and not without considerable controversy. Those who estimate these costs are often criticised that they make no attempt to incorporate into their analyses the positive effects of invasive species.[13] Typically, there *are* some positive effects. Many invasive animal species are predators that eat animals that we consider pests. Wasps are a classic example of such predators. And wasps might even be helping to reduce global warming, as we saw in the previous chapter.

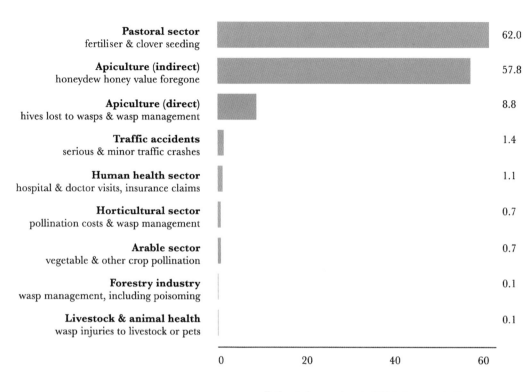

Pastoral sector fertiliser & clover seeding	62.0
Apiculture (indirect) honeydew honey value foregone	57.8
Apiculture (direct) hives lost to wasps & wasp management	8.8
Traffic accidents serious & minor traffic crashes	1.4
Human health sector hospital & doctor visits, insurance claims	1.1
Horticultural sector pollination costs & wasp management	0.7
Arable sector vegetable & other crop pollination	0.7
Forestry industry wasp management, including poisoning	0.1
Livestock & animal health wasp injuries to livestock or pets	0.1

Estimated annual cost of *Vespula* wasps to the
New Zealand economy (millions of dollars)

The estimated annual economic impact of common and German wasps on the New Zealand economy, from a 2015 evaluation. Wasps are estimated to exert the highest cost to the beekeeping industry ($8.8 million from hives lost to wasp raids and in wasp management; $57.8 million due to wasps excluding honey bees from beech forests in the South Island). The projected cost of wasps to New Zealand until 2050 is $1350 million.

Arguably the leader in the field of analysing the economic effects of biological invaders has been David Pimentel from Cornell University. He and his fellow researchers have estimated that the annual global cost of invasive species is US$1.4 trillion, which, in 1998, was around 5% of the global economy.[14] For example, one species, the red imported fire ant (*Solenopsis invicta*), has been estimated to cost the US economy around US$1 billion each year.[15] These ants damage livestock and wildlife and are a public health issue due to stings and resulting anaphylactic shock. The biggest costs of fire ants come from the efforts councils put into managing them in residential areas and because of damage to electrical and communications facilities. Some attempts at fire ant control, well documented on YouTube, include the use of shotguns, petrol, and molten aluminium poured into nests. People are passionate about fire ants in a very similar way to wasps. Firing a shotgun at a fire ant mound almost certainly isn't an especially effective method of control, but it seems to make Mr Texan feel a lot better about life. So far, unfortunately, those shotgun shells and every other control method have only resulted in fire ants and wasps doing just fine.

We have only one economic analysis of wasp invasions, which was recently carried out in New Zealand. Common and German wasps are estimated to cost the New Zealand economy $133 million each year (the current equivalent of about US$94 million per year).[16] Included in this figure is the cost of wasps to the health system, with traffic accidents making up some of that cost. A headline in one New Zealand newspaper in the summer of 2017 read 'Wasp scares woman, woman jumps out of car, car rolls into sea'. The poor woman in question had parked her car on a boat ramp in the coastal town of Raglan to enjoy the view while having a bite to eat. She clearly hadn't quite come to the point of using the handbrake. 'It was horrible,' she is reported to have said, in response to seeing the wasp. Her car ended up completely submerged and drifting out to sea.[17]

Sacrificing a car to avoid a wasp is just one example of how a fear of wasps can cost. The avoidance of wasps, however, is more serious when the car is moving rather than parked on a boat ramp. Due to individuals desperately trying to avoid them, wasps 'cause' $1.4 million worth of traffic accidents per year in New Zealand.[18] Probably very few of the people who were in those accidents were actually stung by wasps. Instead, most are likely to have been distractedly rolling down a window or frantically searching for a rolled-up newspaper, before colliding with another car or something else better avoided.

But every year, New Zealanders – sometimes in cars, at home or work, and numerous other places – are stung by wasps. Most sting victims probably do little more than yelp or curse. And while some of those people visit their doctors briefly, others experience long hospital stays. Stings are estimated to cost us around $1.1 million per year.[19] And as we've seen, some people even die from wasp stings, the emotional cost of which to loved ones cannot be estimated.

Overall, the known and estimable financial costs of wasps from traffic accidents and to human health is a few million dollars, which in the grand scheme of a country's entire economy isn't huge. They rank at 4th and 5th, respectively, on the estimated scale of economic impacts of wasps.[20] Much more substantial are the effects of wasps on New Zealand's beekeeping industries and pastoral sectors.

The biggest cost

The biggest financial cost of wasps comes to us much more indirectly than through car crashes and stings. Wasps raid bee hives to collect honey, and to kill adult and juvenile bees for food. More than half of New Zealand's estimated annual $133 million cost of wasps is due to damage or lost opportunities to the beekeeping industry. A survey in the late 1980s estimated that 8.1–9.4% of bee hives in New Zealand are lost to wasps each year;[21] a more recent survey indicated the annual loss to be around 5%.[22]

Where wasp densities are low, few bees are killed and little money is spent on wasp control. Beekeepers in areas with high wasp densities, however, report that they lose about one third of their hives to wasps each year. How many other businesses would survive if a third of their staff and products were eaten or stolen during the summer? These beekeepers might spend $20 on wasp control per hive each year,[23] which becomes a hefty cost when you consider that some beekeepers have thousands of hives. Despite the beekeepers' best efforts, their hives are often completely destroyed by wasps. Spring populations of bees are particularly affected in areas of the North Island, where German wasp nests successfully survive the relatively mild winters of New Zealand.

A wasp threatens a honey bee guarding the entrance to the bee hive. The wasp
has her mandibles open wide in a clear warning to the honey bee.
Photo: Chris Robins

New Zealand beekeepers I have talked to describe years in which there are
wasp 'plagues'. In years like these, they can lose entire apiaries and even have to
abandon apiary sites to wasps. 'We lose more hives to wasps than to varroa,' one
beekeeper told me. Varroa is a parasitic mite that spreads viruses and destroys
bee hives around the world. 'The aftermath of a wasp attack on an apiary is a
nightmare. They do a lot of damage and we'd absolutely support an eradication
programme.'

I've seen wasps sit, watch and select their prey. They will pounce and
dismember the bee, quickly removing first the head and then the wings, and
sometimes other body parts, before taking the treat back to their nest. In an
all-out assault on an apiary that could contain about 20 hives, wasps seem able
to determine which hive is the weakest and most poorly defended. The wasps
then focus their attack. In an effort to defend themselves, bees crowd their hive
entrance and aggressively bite and swarm at the invading wasps, smothering
them with workers. In Asian countries, if a larger wasp species such as a giant
hornet attacks, the bees crowd around the marauder. With their combined body
heat, they attempt to cook the invader alive. But in New Zealand, wasps often

A wasp has beheaded a honey bee. Wasps will kill adult bees and eat the larvae and the honey. *Photo: Imagebroker*

overwhelm the bees. A raid might last for days, until the bee hive is completely destroyed and cleaned out. The wasps convert the protein from bee larvae into wasp larvae. Then they attack another weak hive – and so on.

It is difficult to gain an accurate assessment of the effects of wasps on the beekeeping industry. We can estimate direct mortality, but what about the more subtle effects? For instance, bees can spend a lot of time and resources guarding their hives against wasp attack. Without doing that guarding, these bees would be off foraging for food or tending to their larvae and pupae. Many people who live near high wasp abundances will be able to sympathise with bees – a family picnic in autumn quickly becomes a Mission Impossible task of fending off hungry wasps; your ham sandwich and chocolate cake are being swamped by a dozen insects, but right now you need to stop young Tommy from drinking that can of fizzy drink into which one of them has crawled . . . That's similar to the sort of stress bees face every day in late summer and autumn, and throughout the year in areas where German and occasionally common wasps successfully overwinter. Exactly how much energy and loss in productivity does this guarding behaviour cost a honey bee hive? We don't know for sure, but the best estimate we have is that *Vespula* wasps directly cost $8.8 million to the New Zealand beekeeping industry each year.[24] That cost comprises money spent on wasp control, replacement of hives, and the lost foraging ability of bees that are forced to spend so much time guarding their hives from wasp attack.

But the indirect costs of invasive social wasps are much more substantial. Wasps dominate and consume nectar and honeydew that bees might otherwise convert into honey. That collection is the most problematic in and around the honeydew beech forests of the South Island. For these forests, the authors of the New Zealand economic survey made several assumptions, such as a production of 3500 kilograms per hectare of dry weight of naturally produced honeydew from beech trees, and that in the absence of wasps, beekeepers could access around 14% of the total area of beech forest area. Overall, they estimated that wasps annually exclude from these beech forests a $57.8 million honey industry.[25] That exclusion, and the loss of resources that could be converted to honey, is considered the second greatest effect of wasps on New Zealand's economy.

Damage to agricultural and horticultural production

When bees are busy being eaten by wasps or attempting to guard their hives from invasion, they're spending much less time pollinating crops and pasture. That reduction in pollination becomes especially important when we consider the plants upon which the economies of agricultural countries like New Zealand are heavily reliant. Clover is one such important plant, as it's a nitrogen-fixing species. All plants depend on nitrogen – without it, they die – but most plants can't use nitrogen as a gas. Clover, however, can draw atmospheric nitrogen from the air and store it in its roots, which improves the quality of the grass that animals are eating, and consequently reduces the need for fertiliser. If wasps suddenly disappeared and we saw a small increase in pollination – just a 2.5% increase, say – we wouldn't need to use as much fertiliser, and we'd see an improvement in the self-seeding success of clover. This small 2.5% increase in pollination would probably result in an increase of 1–3 kilograms per hectare in nitrogen fixation on farms. Due to their effects on clover and the need for nitrogen fertilisers, wasps are estimated to cost the agricultural industry $62 million each year.

Some of the nitrogen fertiliser that farmers spread on their land is washed into streams and waterways, where it becomes yet another environmental concern. The Ministry for the Environment in New Zealand reports that nitrogen levels in our waterways are rising due to intensive agriculture and practices such as the application of nitrogen fertiliser.[26] In the same report, three quarters of New Zealand's monitored native fish, and a third of our invertebrates and plant species were shown as nearing extinction. For Māori, fresh water is taonga (a highly prized object or natural resource) and essential to life and identity. Freshwater ecosystems frequently provide valuable resources and support Māori values and practices including healing and food harvesting. Īnanga (whitebait, of the *Galaxias* species), tuna (freshwater eels), and kōura or kēwai (freshwater crayfish) are treasured. Twenty-eight of the 39 sites assessed by the Ministry for the Environment had a poor or very poor status for mahinga kai (traditional food gathering or cultivation).[27] The presence of wasps and their influence on bees and the subsequent need for fertilisers is just one of the many factors causing the

A wasp feeding on juicy *Coprosma* berries (a native New Zealand plant genus). Wasps have a hugely varied diet – plant nectar and fruit are a source of carbohydrate, and animals are a source of protein. They are especially drawn to ripe fruit from crops and native plants. *Photo: Lucy Johnston*

degradation of waterways. But that small 2.5% increase in bee pollination of clover, which would result if wasps were no longer around, could be a big step in the right direction for these freshwater ecosystems.

Because they affect bees, wasps also influence the pollination in crops of kiwifruit, apples, pears, peaches, apricots, plums, nectarines, cherries, oranges, lemons, mandarins, tangelos, strawberries, boysenberries, blackcurrants, blueberries, feijoas, tamarillos, passionfruit, persimmons, avocado, peas and squash![28] Writing in 1980, entomologist Robin Edwards suggested that poor fruit production in some areas of New Zealand is because wasps kill so many bees that flowers are not pollinated.[29] Throughout New Zealand as a whole, however, wasps in fruit crops are more recently estimated to have much less of a negative effect relative to that seen on pasture land.

South America is experiencing similar effects as a recent population of common wasps in Argentina moves outward into Chile and eventually upward into Bolivia and Peru.[30] The common wasp has been recorded as absent from South Africa, but the German wasp has been present there for more than 40 years. In a 2015 survey by Karla Haupt, most people said they wanted the South African government to initiate an eradication programme, and almost all people (95% of those surveyed) wanted wasps to be removed from their properties. 'I very badly want to get rid of the wasps – soon, quickly, now!' urged one person living with wasps. One grower succinctly stated that the wasps are 'a hell of a big problem'. Wasps here result in crop loss – not only because wasps eat crops, but also because crop harvesting becomes dangerous for the workers. 'Easily 200–300 kg of grapes could not be picked, as the wasps were just too bad,' one farmer said. 'We now have to send a worker the previous day to make sure that *Vespula* is not present in a certain vineyard block'.[31]

For New Zealand, the effects of reduced pollination on agricultural and horticultural sectors equate to tens of millions of dollars per year. We also see effects on forestry and tourism. Tourists don't seem to include wasp stings in their fond memories of New Zealand, and wasps often influence New Zealanders' recreation activities. 'Wasps cause the young people that I take to experience the outdoors to not want to ever return,' said one survey respondent. 'The wasps often cause multiple stings and turn what would have been a totally positive outdoor experience into a negative one. They are a massive annoyance.'

I have Chilean friends who now take a fish when they go for picnics. The fish isn't for human consumption: they place it a little distance from their picnic site, in the hope that the abundant little scavengers will focus on the fish rather than the picnic. The fish is an offering to the wasp gods, in exchange for a little peace at lunchtime.

They aren't all bad . . .

A cost-benefit analysis of the effects of wasps on their invaded range would be incomplete without mentioning potential benefits. Some people, including Ken Thompson, have criticised David Pimentel's work for its heavy focus on the negative impacts of exotic species and for largely ignoring the positive, even vital influence of many introduced species, such as honey bees, on our crop production and economies.[32] Wasps, too, could play some role in pollination of some crops. Wasps could help control pests such as flies and aphids. They may even have had a role in limiting the establishment and spread of another recently arrived species in New Zealand, the great white butterfly (*Pieris brassicae*), which was first observed in 2010. The high abundance and predation rate of wasps may have helped in the successful eradication of this exotic herbivore, though an intensive eradication campaign by some skilled and committed people should primarily be given the credit. Later we'll look at the role of wasps as potential biological control agents, though it's fair to say that with our current knowledge it is impossible to put a dollar figure on these services that wasps might provide.

As we've already seen, at one time wasps were being exported from New Zealand for people to eat. One New Zealand company, led by Geoff Watts, collected and shipped wasp larvae to Japan for years. At peak production, Geoff was exporting 3.5 tonnes of wasp larvae. That's the equivalent of about 12,000 nests. While there is still some demand from Japan and other countries for wasps, those shipments have more recently had to stop, as Geoff found that his business was no longer viable. He told me that there are the obvious, significant health and safety issues associated with collecting wasps. Every day of nest collection, you're dealing with hundreds of thousands, if not millions, of angry wasps determined to inflict harm. In addition to the quarry wanting to hurt his staff, Geoff faced several other major challenges. It's very difficult to find good staff who don't mind an occasional wasp sting. And these wasp collectors needed to wear bulky and protective suits that were so uncomfortably hot that periodically people would faint. The laws that regulate such dangerous jobs have become increasingly strict and costly, so the stings and fainting episodes that were once a bit of a joke now need to be taken much more seriously. There is concern, too, that the wasps in many collecting areas might have been subject to control by pesticides, and, clearly, pesticide residues need to be kept out of food.

All of these issues could likely be overcome if there was massive demand for New Zealand wasps, which would drive prices up and enable collectors to be paid well and profits made by all. However, China exports large quantities of wasps to Japan, and even though New Zealand wasp larvae is rumoured to be of higher quality, it's still not enough to make New Zealand wasp exports economically viable. So currently there are no exports of wasp larvae from New Zealand to Japan.

Today, Geoff Watts runs a pest control business. I asked him: if we were to somehow eradicate or drastically reduce wasps from New Zealand, how would his livelihood change? I'm curious as to whether wasps have the positive effect of keeping some people employed. 'Wasp control does provide some income, especially in late summer and autumn,' he replied. 'If wasps were controlled, we'd have less revenue, sure, but we'd survive.' Pest control companies elsewhere would probably experience more or less of a downturn should numbers of wasps fall. However, Geoff doesn't seem too bothered by the prospect of a wasp-less future.

Do Vespula *wasps exert 'net harm' in their invaded range?*

Thinking back to Ken Thompson and his book *Where Do Camels Belong?*, I find it useful to consider his concept of *net* harm. His wording again: 'We should be certain, from an honest, objective analysis of the best available evidence of its positive and negative impacts, that our intended target is causing *net* harm'. His conclusion was that we should think very carefully about if and how we should manage pests. 'I am saying that we should commence any attempt to control alien species with our eyes wide open.'

The cost of wasps in New Zealand from 2017 until the year 2050 is projected to be $1350 million.[33] Many assumptions were made, and some tested, which enabled the authors to put some confidence intervals around their project. The least they suggest that social wasps might cost New Zealand is $700 million from 2015–2050. The upper limit would be $2 billion, or more.[34] This estimate and my best guess would be that the economic costs of wasps vastly outweigh all positive influences they may have. The positive benefits would likely offset the costs to some extent. Some businesses have been built around invasive social wasps and their control, and it's possible that their economic value could be enhanced. But the economic analysis found that there is insufficient evidence to draw conclusions about the benefits of wasps to the country.[35]

The effects that wasps have on New Zealand's economy and society are only part of their net harm. It's clear that they weigh heavily on our environment. As predators, they put immense pressure on native moths and spiders. We see negative effects on New Zealand's native birds, especially in the honeydew beech forests. And it's impossible to put a value on the lives of people who have been killed by wasps.

I can only conclude that wasps absolutely cause net harm in their invaded range of New Zealand. I'm confident too, to a degree, that they have similar environmental and economic effects elsewhere in their new territories. Argentina, South Africa and Tasmania, for example, are likely to be suffering from wasps.

4. A COCKTAIL OF NASTINESS

The dramatic effects of a tiny amount of poison

'Woman drives off a bridge after "swerving to avoid wasp inside car"'. That was the headline of a 2015 story in an Australian newspaper. And yes, the wasp the woman was swerving to avoid was inside her car. The 25-year-old woman was so panicked that she swerved her car off a 2-metre-high bridge and ended up in a ditch. The car was totalled. She managed to avoid injury, with the exception of the humiliation of the media story (and now a reminder of that humiliation here). There was no mention as to whether the wasp survived the accident.[1]

I'll admit to having a giggle when I read that story. But for all we know the driver was allergic to wasp stings. Allergic reactions can have dire consequences. In a 2012 case in Switzerland, some stings from angry wasps resulted in a death and a conviction for involuntary manslaughter. The unfortunate casualty was a man who'd been having lunch on his balcony. While he and his wife were enjoying their lunch, their neighbour was having a wasp nest removed. Upset about their eviction, a few of the wasps from this nest went looking for someone to blame – and found and stung both the man and his wife. The man went into anaphylactic shock and died two days later. His neighbour was convicted of involuntary manslaughter and negligent physical injury, and was fined about US$9,000. In Switzerland there are no laws on wasp nest removal, but the judges argued that everyone knows wasp stings can be fatal. They criticised the man for removing the nest without checking to see whether anyone was on a neighbouring balcony. His actions were 'careless and in breach of duty'. The death of the neighbour could have been avoided, they ruled, 'had the accused acted more prudently.'[2]

Hives from head to toe

Most of us don't know what a major allergic reaction feels like. A man named Mark Scaggs, a builder, can tell you first-hand. It was early autumn in Coal City, Illinois. Mark was excavating in a subdivision when a wasp stung him on the shoulder. 'It was no big deal,' he said. It was just a single sting. 'After about two to three minutes, I started feeling a little funny. Within five minutes, I started itching head to toe.' Mark knew about the serious consequences of allergic reactions. He was lucky enough to find medical assistance and receive an injection of adrenaline, before being taken to hospital. 'I had welts and hives from head to toe. . . . It all happened eight to 10 minutes, tops, from the initial sting.' One of the biggest concerns with anaphylactic reactions is the person's throat swelling and closing up. Without oxygen, you have just minutes to live. A sting recipient's heart rate and blood pressure can also change dramatically. A healthy person's blood pressure reading should be around 120 over 80; at 60 over 40, Mark's blood pressure was about half of what it should have been. But his quick thinking and a nearby doctor who was able to give him adrenaline medication within minutes of being stung probably saved his life. He now carries an EpiPen wherever he goes, and urges others to do the same.[3]

Most people when stung have a local reaction, with pain and a small welt around the sting site. The pain might last for a few hours. Others have a large local reaction, which typically involves redness, swelling and itching well beyond the sting site. Your entire foot might swell up after being stung on your toe. You'll feel sorry for yourself, but the reaction will recede after a few days or a week. Having a normal local reaction, or even a large local reaction, however, doesn't mean you'll react in the same way the next time you are stung. The most serious outcome is anaphylaxis. Anaphylaxis is defined as a severe allergic reaction to medication, food or venom. Most hospitalisations are due to people being stung by insects or eating foods that are known to cause such a reaction. Like Mark, someone suffering anaphylactic shock might have a widespread rash and experience a low pulse and even shock as their blood pressure drops. They might have an abnormal heart rhythm (arrhythmia) or even, because of a prolonged lack of oxygen, a heart attack. Abdominal cramps are a common symptom. For pregnant women, anaphylaxis can even result in miscarriage.

The symptoms and expression of anaphylaxis after an insect sting vary between adults and children.[4] I suspect that screaming and crying is more common in children than in most adults. Cursing is probably more common in adults (I work with one researcher whom I can hear being stung at least a kilometre away, every time). A cutaneous or skin reaction is most common overall, affecting 80% of both children and adults. Respiratory complaints are common to nearly 50% of reactions in both adults and children. A loss of blood pressure, or hypotension, is common in adults, causing around half of sufferers to lose consciousness, but it's rare for children to lose consciousness. Anaphylactic shock is probably more likely if you receive multiple stings at one time, or perhaps if you're unlucky enough to have suffered repeated stings in one summer. Mark Scaggs, for example, had been stung earlier that summer without any major consequence. In contrast to food anaphylaxis, a slow onset of the sting reaction means that it is less likely to be life-threatening.

These extreme reactions are caused by a tiny fraction of a droplet of venom. It's estimated that, depending on the wasp species, the average amount of venom you'll receive from a sting is between less than 0.5 microlitres and 2.5 microlitres. (Keep in mind that 1 microlitre is equal to 0.001 millilitres.) When you're stung, it's likely you'll be stung more than once. So you might receive 5 microlitres when an angry wasp stings you multiple times. Let's put that in perspective: an average human man weighing 70 kilograms is around 60% water. A newborn baby carries a bit more water (around 75%) and an elderly person a bit less (around 50%). That means our 70-kilogram man is carrying 42 litres of water. The 5 microlitres of wasp venom represent 0.00000012% of the fluid in this man. Yet those 5 microlitres, or even less, can kill him.

So why does your body go into anaphylactic shock from such a pathetically small amount of venom? How is it that you can die from a sting that contains about as much venom as would comfortably sit on the head of a pin?

Pain

Wasp venom is a cocktail of nastiness. There are multiple chemicals in those 5 microlitres. Each has a different function and role.

Your first sensation upon being stung is . . . stinging. A wasp sting hurts. That pain isn't just because of a little needle being poked into you. The sting contains the neurotransmitters acetylcholine and serotonin. Serotonin, a hormone widely found in plants, animals and bacteria, is located mostly in the intestine – not the brain, as is commonly thought. Serotonin plays a major role in regulating gastrointestinal function. Seeds from many plants also contain serotonin, which may be an evolutionary feature that enables the seeds to pass through the digestive tract quickly and in better condition for germination. Pathogenic amoebae also produce serotonin, so when you have an upset stomach you'll probably have

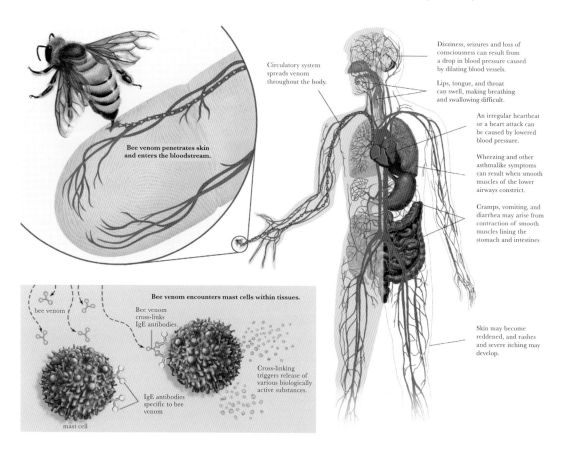

Circulatory system spreads venom throughout the body.

Bee venom penetrates skin and enters the bloodstream.

Dizziness, seizures and loss of consciousness can result from a drop in blood pressure caused by dilating blood vessels.

Lips, tongue, and throat can swell, making breathing and swallowing difficult.

An irregular heartbeat or a heart attack can be caused by lowered blood pressure.

Wheezing and other asthmalike symptoms can result when smooth muscles of the lower airways constrict.

Cramps, vomiting, and diarrhea may arise from contraction of smooth muscles lining the stomach and intestines

Skin may become reddened, and rashes and severe itching may develop.

Bee venom encounters mast cells within tissues.

bee venom

Bee venom cross-links IgE antibodies.

Cross-linking triggers release of various biologically active substances.

IgE antibodies specific to bee venom

mast cell

Systemic anaphylactic response to venom from an insect sting.
Universal Images Group North America LLC / Alamy

diarrhoea, which is the amoebae's way of distributing itself widely and frequently. As well as working in your gut, serotonin can function to promote nerve firing, amplifying the pain you're experiencing from the site of a wasp sting. Serotonin can also work as a vasoconstrictor, which increases blood-clotting. The reduction in blood flow to a sting site means that your blood, carrying your body's defences and venom dilution factors, can't reach the wound easily.

The acetylcholine in wasp venom is also a neurotransmitter – a chemical released by your nerve cells to send signals to other cells. Like serotonin, acetylcholine is produced by a wide array of animals, plants and bacteria, but it appears to have evolved to cause harm, often by hyperactivating muscles and nerves. It can cause an intense, localised, vascular spasm. Acetylcholine also plays a key role in alertness, attention, memory and learning. Perhaps it's acetylcholine that helps you remember just how much that wasp sting hurts. These neurotransmitters, which many animals have, first serve to initiate and amplify the pain of a wasp's little sting.

Serotonin and acetylcholine make up only around 5% of the dry weight of the venom. Bee venom has a small polypeptide (part of a protein) called apamin, which affects your nervous system as well. There is good evidence that it can enhance memory and learning.[5] Mice that have been treated with apamin, for example, show faster learning of how to swim through and navigate water mazes. There is evidence that apamin can help to reverse amnesia and improve your 'fear memory': it may even have evolved in bees to help you remember to avoid bees and their hives in the future. The venom of wasps, bees, snakes and many other animals also contains enzymes called phospholipases and a peptide called mastoparan. Together, these cause the release of arachidonic acid, which causes pain, and the breakdown of mast cells, which is the start of a major problem. A little bit of mast cell disintegration is caused by the wasp peptides. But a lot of the potentially devastating mast cell action is from the sting-recipient's own immune system. In other words, it's your own reaction to the sting, rather than the sting itself, that breaks down the mast cells.

Mast cells, a type of white blood cell, are a key part of your immune system. They're crucial in the healing of wounds and defence against pathogens. They are everywhere in your body where blood flows and are especially abundant where there are boundaries between your body and the outside world – the places that might be exposed to pathogens and parasites, such as the nose and lungs, the digestive tract (including your mouth), and your skin, where wasps typically plant their painful little stingers.

Wasp stings will 'activate' your mast cells. These little immune cells are packed full of granules of chemicals such as histamine and heparin, which are released upon activation. Histamine ($C^5H^9N^3$) is a minuscule chemical, with only 17 atoms. But for a small chemical it does a lot. It is known to have 23 different roles in your physiology. And those roles are quite diverse and diverse, and depend on what type of receptor is targeted. To achieve these 23 different roles there are four different types of histamine receptors on or in different cells of your body.

Type 1 histamine receptors play a big role in your central nervous system, influencing your body temperature and sleep–wake cycle. Type 2 histamine receptors influence your gastric acid secretion. Consequently, these type 2 receptors have become a target for the medical treatment of gastric ulcers. If you inhibit the ability of stomach cells to respond to histamine, then you inhibit excessive acid secretion. Type 2 receptors also have a role in the male erection. One study injected the subjects with histamine intracavernously, which, put more simply, is an injection into the base of the penis. Amazingly, a quarter of the male injectees attained an erection and even more of them experienced at least a partial enlarging response.[6] (Histamine clearly plays a big role here, given the usually shrivelling effect of a sharp needle presented anywhere near that part of the male anatomy.) Type 3, and the most recently discovered type 4 histamine receptors, seem to be the important ones for your wasp stings. In the year 2000, a Japanese team announced the discovery of the type 4 receptors, which has come to represent a major breakthrough in our understanding of histamine and potential treatments for anaphylactic responses.[7] When researchers have experimented with blocking the types 3 and 4 receptors, they've seen a much milder reaction.

The fact that a histamine injection could result in an erection gives a hint as to one of histamine's more common effects. It's a vasodilatory chemical, which means that it facilitates blood flow by relaxing smooth muscle cells within arteries and veins. The intracavernous injection of histamine described above resulted in vasodilation and blood flow into the penis that facilitated an erection. It normally occurs just around those organs where you especially need blood. If your body perceives any kind of threat to your survival – whether that's a literal sharp-toothed predator chasing you, a group of people jumping out and screaming 'Surprise! Happy birthday!' in your living room, or just a suspenseful scene in a horror movie – the arteries and veins around your muscles will vasodilate and allow your muscles to receive a good amount of oxygen-rich blood so that you're ready for action. At the same time, the arteries and veins of your digestive system undergo vasoconstriction: they become narrower, reducing the blood flow there. Your overall reaction to the threat is some selective vasoconstriction and some selective, organ-specific vasodilation. Hence, your overall blood pressure changes but doesn't drop to dangerous levels. However, an anaphylactic, body-wide reaction to a few microlitres of wasp venom results in a huge histamine release with widespread and non-organ-specific vasodilation. That systemic vasodilation makes your blood pressure drop and reduces oxygen flow to your heart and brain. You might then lose consciousness and suffer significant heart damage, leading to death.

The extent of mast cell activation determines the extent of your reaction to wasp venom. The most sensible procedure for your body to follow is to have a piecemeal degranulation of the mast cells and a localised release of the histamine with other inflammatory granules. That results in a normal local reaction, where you might get a small pustule around the sting site, or a large local reaction, where your hand or arm balloons because of a sting to your thumb. Unfortunately, you don't get to decide if you'll have a sensible reaction to a wasp sting. The alternative, and what you don't want to have happen, is a widespread overreaction of your mast cells – an anaphylactic degranulation or rapid histamine release. Each mast cell has a miniscule amount of chemicals like histamine: 2–5 picograms (1 picogram is 1 trillionth of a gram). But the activation and rapid release of histamine by millions of mast cells, with their resulting vasodilation and additional effects, can kill you.

The trigger for the activation of mast cells involves another three key components of your immune system, which serve to ready or prime the mast cells for their degranulation action. You have many types of immune cells. For mammals, one of the important are plasma cells, or plasma B cells. These cells start off the process that helps your body 'remember' any previous exposure to a foreign entity. That entity might be a part of a pollen granule or a component of wasp venom. In response to venom or pollen exposure, the plasma cells make large quantities of antibodies, more scientifically known as immunoglobulin E (IgE) proteins. These antibodies are essentially a lock that fits only a very special key, the key being more of that same foreign entity or something similar to it. The antibodies then attach to mast cells. After the immunoglobulin attachment, these mast cells are ready for action. If a venom component attaches onto them in the future, the mast cells release their load of histamine and other chemicals. If that release happens just in your thumb, you are okay. If mast cells throughout your body release en masse, then you have a problem: anaphylaxis.

'An extraordinary phenomenon'

The formal discovery and science of anaphylaxis goes back to 1901, with two French physiologists named Paul Portier and Charles Richet, and their dogs Galathée and Neptune.[8] Unfortunately this story doesn't turn out well for either of the poor mutts.

It was Prince Albert the First of Monaco who asked the two researchers to investigate the toxicity of jellyfish. Prince Albert was a dedicated oceanographer who spent considerable time at sea; he probably had first-hand experience of venom from a jellyfish tentacle or two. He invited Portier and Richet onto his boat, which had a fully equipped laboratory and animal 'test subjects' with which to study the properties of the venom from the Portuguese man-of-war. This jellyfish has an excruciatingly painful sting, though it is seldom deadly. After returning to land, the scientists continued their work on toxicity at the University of Paris. In an attempt to provide the dogs with some degree of tolerance, the researchers injected Galathée and Neptune with a toxin from sea anemones. Essentially, this was an attempted vaccination. After their first injection, Galathée and Neptune experienced a little itching and some heavy breathing. They weren't pleased about their treatment but were otherwise fine. But neither dog was fortunate the next time. This second injection of the same venom resulted in what the researchers termed 'aphylaxis', a word derived from the Greek *a* (contrary to) and *phylaxis* (protection). The dogs lost consciousness and had the abdominal cramps we see in human anaphylaxis that resulted in vomiting and diarrhoea, and finally both Galathée and Neptune went into a stupor and died. The dogs' treatment by the researchers sounds cruel, but in their defence the physiologists had believed that the dogs would be protected by their previous exposure to the toxin. They were surprised by the fatal reactions after the second injection.[9]

The term 'aphylaxis' was later changed to 'anaphylaxis', because, well, it sounds better and is easier to pronounce: anaphylaxis has euphony, the quality of being pleasing to the ear. Charles (not Paul, and certainly not Galathée or Neptune) won a Nobel Prize in Physiology for Medicine for these studies in 1913. In his acceptance speech for the prize, his sense of wonder is clear. 'It is an extraordinary phenomenon that so insignificant a quantity of poison can modify the organism to the extent that the succeeding days down long years can not eradicate this indelible modification.' What we now know is that the dogs had almost certainly developed antibodies to a part of the anemone venom. These antibodies attached themselves to mast cells, and released histamine after the second exposure. And yes, this early work shows that other mammals and vertebrates can suffer anaphylactic shock. As far back as 1913, as noted then by Richet, 'Anaphylaxis has been observed in all animals: the horse, the goat, the ox, the rat, the pigeon, the duck and even recently in frogs'.[10] You might now see catteries and kennels advertising that they are medically prepared to treat such reactions.

If you are stung and go into anaphylactic shock, it might not be over as quickly as you'd hope. A little less than a quarter of sting-related anaphylactic shock recipients experience biphasic anaphylaxis. That's when you have an initial immune response, a period of no symptoms for a few hours, and the subsequent recurrence of anaphylaxis despite no further stings. If you are really unlucky, you might experience protracted anaphylaxis. This is an anaphylactic reaction that can last days or even weeks, perhaps without ever resolving completely. You (or your cat or dog) can even experience non-immunologic anaphylaxis. This still involves a massive mast cell degranulation, but it occurs in the absence of antibodies that trigger the immunoglobulins. Non-immunologic anaphylaxis can be caused by exposure to drugs such as opiates. Luckily, pets aren't frequently exposed to the common opiate painkillers that people take and that you probably have in your medicine cabinet, like codeine.

Prevalence

It's difficult to estimate just how prevalent anaphylactic shock is, given that its symptoms and treatments are similar to those for heart attacks and other ailments. One recent study in the United States estimated that 5.1% of the population *probably* had experienced anaphylaxis. The researchers estimated that 1.6% had *very likely* experienced anaphylaxis.[11] So, for an adult population, this means that between 1 in 50 and 1 in 20 people will suffer a life-threatening anaphylactic shock at some point in their life.

Estimates of anaphylaxis specific to insect stings are that they occur in about 3% of adults.[12] But we can't say for sure how many of those cases are due to wasps, because, as we know, people frequently can't tell the difference between a wasp and a bee. Perhaps the best data comes from Europe, which indicates that around 20% of all anaphylactic shock cases were due to insect venom. Of those, 75% were due to wasps or hornets, and about 23% from honey bees.[13] Hospital admissions for anaphylaxis are increasing by around 11% per year in England, and are near that rate in much of the rest of the world. So what's the mortality rate due to hymenopteran stings associated with anaphylactic shock? As we saw earlier, the most recent estimate of the mortality rate from wasps in New Zealand is approximately two deaths per year, for a population of 4.4 million people.[14] In Australia, a recent analysis showed that hymenopteran stings are the highest cause of hospitalisation and the first-equal cause of mortality. Ant, bee and wasp stings have killed more Australians than have sharks and crocodiles combined over the period of 2000–13.[15]

Over time, we've been seeing many more wasps and many more people being stung. In the scientific community, the strong consensus is that climate change has a strong bearing on this change. For example, over the last 50 years the average temperature in Alaska has risen by 2.2°C. As wasps appear to survive the milder climate, there has been a sevenfold increase in stings in northern Alaska between 1999 and 2006.[16] In 2006, two people in the city of Fairbanks died after going into anaphylactic shock from wasp stings. This was the first time Fairbanks had reported such deaths.

As in Australia, in the United States the most common animal phobias are spiders and snakes. But looking at it rationally, just how common are bites from spiders and snakes? Science journalist Rachel Nuwer looked at the database provided by the Centers for Disease Control, and found that between 1999 and 2007 in the United States, where a total of 714 people died after being bitten or stung by a venomous animal, only 59 people died after being bitten by a snake or lizard. Spiders caused permanent 'lights out' for just 70 people. By comparison, wasp, hornet, and bee stings were the cause of 509 of those 714 deaths.[17] Many of those stings are likely to have been from honey bees, as many people love honey bees and fear them less. But they and their wasp cousins are significantly more likely to kill you than your average rattler. Of those deaths from bees and wasps, 81% were men or boys, which brings to mind the 'wasp sting stupidity chart' that we saw earlier. New Zealand males typically rated highly on this stupidity scale, receiving 75% of the stings overall. Rachel likewise surmised for the US data that 'men seem to be universally prone to such needlessly risky, even dumb behaviour.' Fair enough: it is hard to argue with the data.

Life savers

All this doom and gloom is a bit depressing. But there is cause for hope: you can become desensitised to venom. The first documented example of desensitisation dates back to 1925, with a doctor in Cape Town, South Africa, working with bee venom. A 32-year-old woman, who 'live[d] in a district of wild bees', would clearly go into anaphylactic shock after receiving a sting and would require adrenaline as treatment. For the immunotherapy, her doctor applied diluted and increasing concentrations of venom every two days to the 'sacrificed area', until the woman could tolerate injections of undiluted bee venom. In a final, somewhat risky test, 'she then heroically suffered herself to be stung by a bee'. Ah, the good old days, when the ethics and the question of a person's or a dog's survival were in a distant second place to the advancement of science. Fortunately, the patient's reaction was described as 'extremely mild'.[18]

This process is known as venom immunotherapy (VIT), and it's generally an extremely effective way to desensitise yourself against insect venoms. The process hasn't really changed much since 1925. It involves the delivery of a small amount of venom via a skin prick test to determine your reactivity. Then, every few weeks, you are injected with extremely low concentrations of venom. This process can continue for up to five years. An initial dose for the treatment of bee stings might start with an injection of approximately 0.0001 micrograms and rise to 100 micrograms, which is the equivalent of around two bee stings, or 30–50 individual wasp stings. The initial response is that your mast cells become desensitised, followed by an array of changes to your immune system.

Most studies on VIT show that it can dramatically reduce the incidence of non-fatal venom anaphylaxis as well as the risk of fatal anaphylaxis. People receiving this treatment often report a 'more positive view of life', too, which is understandable when considering that an allergy to insect venom can cause such distress. The use of a live insect – typically a bee – to deliver the final test actually seems to be of benefit too, probably because people view it as a more realistic, 'real-world' test. So it does seem that our experimenting doctor in South Africa made the right call. But, even though we know that VIT is effective, it's expensive and can be time-consuming. In Britain it's reported to cost £2,300 per patient.[19] Because of this, it's more recommended to people who are at risk of multiple stings over a short period, such as bee keepers.

If you know you are allergic to wasps and you don't undergo venom immunotherapy, you should carry adrenaline. Mark Scaggs' life was almost certainly saved by his ability to access adrenaline; since then, he's recommended that people carry auto-injectors in case of anaphylactic shock. So how does adrenaline work? As we saw earlier, the life-threatening situation of a hungry predator with sharp teeth and scary claws can lead your body to experience some selective vasoconstriction in organs like your gut, and some vasodilation in your leg muscles. Over millions of years, humans have evolved this response to push oxygen-rich blood to the places where it is needed the most, rather than to places where it isn't currently essential to your immediate survival. Those effects are largely due to adrenaline: the fight-or-flight hormone. Today, obviously sabre-toothed predators don't lurk around places where most people live, but the legacy of these fears is apparent when, for instance, your boss sends you an e-mail asking for an urgent meeting, or when a wasp suddenly appears inside your car and you are sufficiently alarmed to swerve off a bridge in an attempt to avoid it.

Adrenaline increases your heart rate and blood pressure. This results in the expansion of the air passages in your lungs, which are swelling due to the effects of histamine. Adrenaline redistributes the blood, closing some arteries and capillaries and opening others. It can change and increase your blood glucose levels, increasing the sugars and energy in your brain. The downside of adrenaline as a treatment for an allergic reaction is that it doesn't last long. If you're stung, you might need multiple doses of adrenaline over many hours. Most of us on the receiving end of an insect sting won't need adrenaline, but an effective medication to relieve the itching and pain will always be welcomed. Something to reduce the itch and overwhelming need to scratch.

Experiments with itchy mice have only recently helped us understand why scratching an itch only makes it worse.[20] Histamine is the reason for your desperation to scratch that sting site. One of the effects of histamine is to cause your nerves to send a please-scratch-me signal to your brain. Your nerves that send itch signals are intertwined with the nerves associated with pain signals. When you are either overwhelmed by the need to scratch, or are distracted and forget to do so, the itch causes a small degree of pain. The little bit of pain caused by scratching relieves the itch, momentarily. And it feels so good! Until, unfortunately, your brain releases serotonin in an attempt to dampen that pain signal. That serotonin travels from your brain to your nerve cord. The serotonin has the unfortunate side effect of generating further itch signals. Scratching can also inflame tissue and cause more histamine release. And so the feedback loop begins: more scratching leads to a little pain, some serotonin release, and more please-scratch-me signals, and more scratching and histamine release. The best, though seemingly impossible, advice for any sting that causes a local reaction is to resist all temptation to itch!

In an attempt to limit your histamine release you can swallow some antihistamine tablets. Or rub some antihistamine cream on the offending area. These treatments can be useful for a normal local reaction, especially within one or two hours after being stung. Antihistamines might help to lessen the please-scratch-me signals and break the feedback loop. But antihistamines are not recommended for treating anaphylaxis, and for a lot of people they don't seem to do much. Some people state that rubbing toothpaste on the sting site is more effective than taking antihistamines. Another oft-repeated but useless remedy for stings, especially jellyfish stings, is to have someone urinate on the sting site. Unfortunately, having someone take a whizz on you could make the problem worse. If there are jellyfish tentacles still attached to you, the urine might aggravate the tentacles into further venom delivery. And there is no evidence that urine provides any pain relief. Should a 'friend' ever offer to apply this special treatment to some part of your stung anatomy, you should reconsider both their motivation and your friendship.

'A hot, red, enduring flare'

How much does a wasp sting hurt? Because it is subjective, pain is an experience that is hard to quantify and compare. I once dislocated my shoulder playing rugby, and for a decade it would occasionally pop out of its joint, until I had surgery. I still writhe in sympathetic pain whenever a dislocation occurs during the All Blacks' annual defeat of the Australian national rugby team. I've been told that a shoulder dislocation is more painful than childbirth. But having watched my wife have our two boys I'm not convinced and am extremely glad not to have the personal opportunity to directly make that comparison.

An entomologist named Justin Schmidt has spent a lifetime comparing and classifying the painfulness of various hymenopteran stings. Over the course of his career, he has travelled all over the world in order to be stung by more than 83 different species of insect, in the name of science. It's not that he wants to be stung, precisely; instead he wants to know what it *feels like* to have been stung. He is therefore known as the 'Connoisseur of Pain' and 'The King of Sting'. In order to classify the painfulness of these stings he developed and refined the Schmidt Sting Pain Index, a four-point-star sting pain scale for hymenoptera stings. The numeric ratings are accompanied by specific descriptions. The lowest-ranked stings are the likes of the anthophorid bee, at level 1: 'Almost pleasant, a lover just bit your earlobe a little too hard'. Of wasps, the lowest are species like the club-horned wasp, at 0.5: 'disappointing; a paperclip falls on your bare foot'. I've never met a club-horned wasp. But I'm now no more scared of them than I am the tub of stationery in my desk drawer.

Other wasps, however, can pack a punch, according to the Schmidt Sting Pain Index. The most painful wasps include the tarantula hawk wasp, with the highest possible score of 4: 'Blinding, fierce, shockingly electric. A running hair dryer has just been dropped into your bubble bath.'[21] Tarantula hawks are common throughout much of the United States. The best thing you can do after a tarantula hawk sting, Justin advises in his paper 'Venom and the Good Life in Tarantula Hawks (Hymenoptera: Pompilidae): How to Eat, Not be Eaten, and Live Long', is to lie down and start screaming, because 'few, if any, can maintain normal coordination or cognitive control to prevent accidental injury. Screaming is a satisfying expression that helps reduce attention to the pain of the sting itself.'[22] He recounts how one quite famous entomologist, Howard Evans, reacted after being stung while attempting to retrieve some wasps he'd captured in an insect net: 'Undeterred after the first sting, he continued, receiving several more stings, until the pain was so great he lost all of them and crawled into a ditch and just bawled his eyes out.'[23]

The only *Vespula* wasp on Justin's list of 'things he has been stung by' is the Western yellowjacket, *Vespula pensylvania*.* The wasp rated a 2 on the pain scale, about the same rating as a honey bee's sting. It is described as 'hot and smoky, almost irreverent. Imagine W.C. Fields extinguishing a cigar on your tongue'. These stings produce

> instantaneous, hot, burning, complex pain that gets one's attention no matter what other thoughts were preoccupying the mind. The pain lasts unabated for about 2 minutes, after which it decreases gradually over the next couple of minutes, leaving us with a hot, red, enduring flare to remind us of the event in case our memory should fade. These stings are worthy of storytelling to loved ones.[24]

My experience with common wasps and German wasps is a little different, and I can only provide a much less imaginative description. To me, both wasp stings feel like someone is pushing a hot, sharp thumbtack into my skin. The burning declines over the next hour, but I'll feel these stings for at least a week. Ten days later, the sting site will suddenly itch and remind me to be more cautious. Of course, the experience of stings, and of pain generally, differs for

---

* Those of you familiar with the spelling of the state of Pennsylvania might wonder about the wasp's origin and come to the realisation that the entomologist who named this wasp couldn't spell. He also got the actual geographic origin of the wasp wrong.

everyone. The amount of pain you'll experience also depends on where on the body you might be stung. Take a minute to think about the part of the body where you'd imagine a sting would be the most painful.

Whatever you're thinking, you're probably wrong.

While he was a graduate student at Cornell University, entomologist Michael Smith had the bright idea of carrying out a comparative study to determine the worst part of the body to be stung by a bee. 'This study,' he wrote in his abstract, 'rated the painfulness of honey bee stings over 25 body locations in one subject (the author).'[25] He was working with honey bees, not wasps, but I imagine the results are similar and, personally, I am not willing to repeat his experiments with wasps just to check. His experiment lasted 38 days. You might wonder about the ethics of such experiments. Well, Cornell University was perfectly willing for Michael to inflict upon himself what turned out to be an excruciating and eye-watering amount of pain. Michael also notes that the cornerstone body on human research ethics, the Helsinki Declaration of 1975, was perfectly happy with his experimental self-harm. So he examined a total of 25 different locations throughout his body. Honey bees were held against toes, fingers, hands, feet, legs. The top of his skull was stung, as were his lip, nostril, nipple, scrotum and penis. And not just once; the experiment required that all body parts endure three replicate stings. Michael carefully selected the most aggressive bees from hives for this purpose.

He rated the pain on a scale of 1 to 10, though he notes that 'all the stings induced pain in the author'. The three least painful locations were 'the skull, middle toe tip, and upper arm (all scoring a 2.3)', while the three most painful locations were 'the nostril, upper lip, and penis shaft (scoring 9.0, 8.7, and 7.3 out of 10, respectively).[26] So stings to the nostril win the competition! I asked Michael to describe the experience and how much he must have looked forward to replicating the nostril stinging.

A honey bee sting to the nostril is a whole body experience. Your eyes tear up, your nose is spewing mucus, you're sneezing, and all you want to do is get that stinger out (I had to keep it in there for a minute, which was unpleasant, but necessary). And you are correct, when I decided to do the third replicate, that was the only location that I was dreading to repeat.

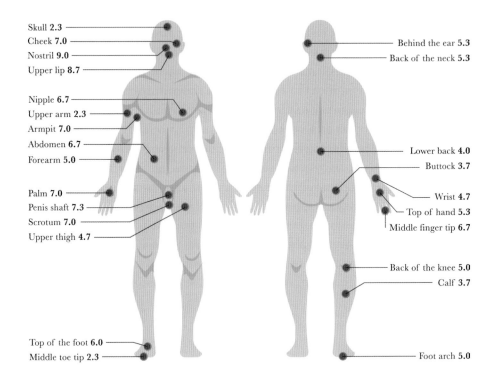

Skull **2.3**
Cheek **7.0**
Nostril **9.0**
Upper lip **8.7**

Nipple **6.7**
Upper arm **2.3**
Armpit **7.0**
Abdomen **6.7**
Forearm **5.0**

Palm **7.0**
Penis shaft **7.3**
Scrotum **7.0**
Upper thigh **4.7**

Top of the foot **6.0**
Middle toe tip **2.3**

Behind the ear **5.3**
Back of the neck **5.3**

Lower back **4.0**
Buttock **3.7**
Wrist **4.7**
Top of hand **5.3**
Middle finger tip **6.7**

Back of the knee **5.0**
Calf **3.7**

Foot arch **5.0**

The dots show the 25 locations of repeated honey bee stings that Michael Smith applied to his body over a 38-day period. The scores represent pain ratings. Low numbers represent little pain; high numbers represent great pain. No one has yet repeated this experiment with wasp stings (I certainly won't).
Data and figure adapted from Smith (2014)

Michael also noted that, if you are forced to choose between a sting to the nose or to the penis, 'you're going to want more stings to the penis.'[27]

Deservedly, Michael won a major international prize for his work. The 2015 Ig Nobel Prize in Physiology and Entomology was jointly awarded to him for this study and to Justin Schmidt for his Sting Pain Index.[28] The Ig Nobel prizes, for 'research that makes people laugh and then think', are handed out annually by 'genuinely bemused genuine Nobel laureates'. Other 2015 Ig Nobel winners included a group of biologists for 'observing that when you attach a weighted stick to the rear end of a chicken, the chicken then walks in a manner similar to that in which dinosaurs are thought to have walked'.[29] That was a serious scientific study, but as with both Michael and Justin, the authors could see the funny side and happily accepted their awards.*

Wasp venom as zombie-maker and curative

My favourite interesting wasp and wasp venom story is that of the emerald cockroach wasp, also known as the jewel wasp (*Ampulex compressa*). The jewel wasp is a beautiful emerald-green wasp that hunts adult cockroaches. It hunts them not to enjoy for itself but to use as live prey for its young larva to feast upon. But there is a problem. The wasp needs its cockroach, which is larger than the wasp, to stay in one place and not move. This way, the wasp larva can eat it at its leisure. The wasp has somehow evolved some fantastic venom for this purpose.

On first encounter, the jewel wasp delivers a sting to the cockroach body, then steps back. The sting temporarily paralyses the cockroach and allows the wasp to deliver a more precise sting. Sting number two is very carefully delivered to a specific part of the cockroach's brain: the part that controls the escape reflex. Again, the wasp steps back. The cockroach appears to recover and groom itself, though it does not and cannot move away. The venom seems to work by blocking the action of a neurotransmitter called octopamine, which seems to help rob the cockroach of its free will.[30] The wasp breaks one of the cockroach's antennae and seems to taste its haemolymph (insect blood). It is thought that the wasp may be tasting the blood to determine the concentration of venom in it. The jewel wasp takes the cockroach by the other antennae and leads it to a hole in the ground. There, it lays an egg on the cockroach. The egg hatches. And the larva that emerges slowly devours the living, motionless cockroach alive. The

---

* The most bizarre Ig Nobel I've ever read about was a public health prize award for 'Surgical Management of an Epidemic of Penile Amputations in Siam' – techniques that are recommended for reattachment, provided that the amputated penis had not been 'mutilated, decomposed, or partially eaten by a duck.' That one was so strange I had to find and read the publication in question. I wish I hadn't. It has pictures.

non-essential cockroach organs are eaten first. Finally, the essential organs are consumed and the new wasp pupates inside the shell of the dead cockroach body. I'm sure that this type of parasitoid–host interaction inspired the creators of the *Alien* movies.

That's just one of many stories about how venom can influence behaviour. In this world we have mind-controlling wasps that use their venom to enslave zombie spiders to build protective shelters. We have wasps that inject virus-like particles in their venom, along with their eggs, into caterpillars, seeming to cause a dying caterpillar to protect the larvae that are effectively killing it.

You might loathe the effects of venom, but it turns out that it can have some extraordinarily interesting, even beneficial effects. A Brazilian social wasp by the name of *Polybia paulista* produces in its venom a small peptide that may help in the fight against cancer, by killing cancer cells while leaving healthy cells unharmed.[31] This venom peptide, or mastoparan, is a small grouping of amino acids. This one from the Brazilian wasp has been called Polybia-MP1, and its effects are specific to cells that have turned cancerous. Unlike healthy cells, cancer cells have abnormalities on their cell walls: the lipids or little fats that help make up these walls are distributed differently compared with every other cell in your body. By a purely random happening in biology, Polybia-MP1 interacts with that abnormal lipid distribution found in cancer cells. It creates a hole in their cell wall, causing crucial cell components to leak through the hole. After this leakage, the cell dies. Polybia-MP1 is also an anti-microbial agent, so it probably acts in a similar way with bacterial cells.

Venom from other wasps can be useful too. Even venom from the stings from our common wasp could be beneficial. *Vespula vulgaris* produces a mastoparan named V1 (MP-V1), which has considerable potential as an antimicrobial agent; some authors even suggest it has superior antimicrobial activities.[32] It shows activity against pathogenic bacteria such as *Streptococcus* and *Salmonella*, and fungi. And, crucially, the bacteria are killed but the mastoparan appears not to harm healthy human cells.

Venom can do so much more than just sting.

5. 'WE NEED TO REDUCE THEIR POPULATIONS'

Options for wasp control or eradication

As a kid, I used to follow my father after dark with a bottle of petrol. He'd spotted the wasp nest entrance earlier in the day. He'd upend the bottle and jam it into the nest entrance, then we'd make a swift retreat. Positioning the bottle near the entrance would allow the petrol to permeate the nest. Hopefully the fumes would kill off any additional wasps and their larvae not in direct contact with the liquid itself.

This was the standard method for wasp control for most people in New Zealand and in many places around the world. People still use it today. Occasionally, people think it is also a good idea to ignite the petrol. A high level of resentment towards wasps means that many people like the idea of burning entire nests. Given that nests are typically found in late summer or early autumn, when the environment is dry, warm and flammable, putting a match to that petrol can mean watching crop fields, tractors, hay sheds, houses and much more go up in flames.

Even though soaking the ground with petrol is risky and environmentally not-so-friendly, using petrol plus fire is the law for social insect control in many countries. Not for controlling wasps, but for dealing with honey bees that are afflicted by disease. Among the honey bee diseases, American foulbrood is one of the worst. Governments around the world have instigated legislation for control or eradication programmes, requiring beekeepers to inspect their hives for this bacterial disease. If they find foulbrood, the beekeepers are required by law to kill the hive by pouring petrol into the nest boxes. The hive must then be burnt. But not immediately, as this can result in explosions, especially if the hive is full of energy-rich wax and honey. I've seen amateur beekeepers devastated and in tears after having to destroy their beloved bees and hives in this fashion.

The aggressive, stinging behaviour of wasps motivates an understandable desire in many people to burn wasp nests, and to draw attention to themselves doing so. YouTube is filled with exceptional, average and 'other' clips documenting unconventional wasp control techniques. Amongst the exceptional is the Chinese military using a flamethrower to destroy a large nest of the Asian giant hornet (*Vespa mandarinia*).[1] This 5-centimetre-long hornet, known colloquially as the yak-killer hornet, regularly kills people, as well as the occasional large ungulate. This giant wasp is a serious problem if it is in your backyard. Among the category of 'other' videos are a number of unfortunately instructional videos on setting fire to wasp nests using an aerosol can and a match. Some of these videos demonstrate the pain and lack of success that can result. I'm sure even worse outcomes don't make it onto the internet.

The Chinese military use a flamethrower to destroy a nest of the Asian giant hornet (*Vespa mandarinia*) high in a tree. For some, this approach is the dream method of dealing with social wasps. *Image: YouTube*

Anyone who takes wasp control into their own hands only to cause a house fire is commonly told to 'leave it to the professionals', as we heard in this report:

> A man who set fire to his house while burning a wasp nest with a spray can and a lighter is lucky he didn't burn the building down, a firefighter says. The man tried to get rid of the nest using the makeshift blowtorch through a hole in the wall to 'smoke the wasps out'. Instead he started a fire in the wall cavity.[2]

Firefighters typically express disbelief at these kinds of situations. 'It's a good time to remember that aerosol cans and lighters are not a safe approach.' Such failures aren't limited to New Zealand. From the UK's *Telegraph*: 'Man sets fire to neighbour's house trying to smoke out wasp nest'. This was the story of a young mother and her two daughters who were left homeless after a neighbour accidentally set fire to their house.[3]

Apparently, if you are in the right country and wait long enough, wasp nests might even spontaneously combust. In 1878, a letter-writer to the journal *Nature* reported on the spontaneous combustion of a large wasp nest under the roof of a house in Port-au-Spain, Trinidad, nearly setting the house on fire. The day had been 'exceedingly hot'.[4] Perhaps with global warming we will see an increase in the frequency of spontaneous wasp nest combustion. As a method of wasp control, fire is hazardous, especially if you are dull enough to light the nests yourself or wait around for their spontaneous combustion. So what else can you do?

From the Middle Ages up until the 19th century, problematic insects were threatened with excommunication from the church, banishment, and even execution. While larger animals such as pigs, foxes, wolves, goats and donkeys could be arrested, tried, and convicted,

> pestiferous creatures such as bugs, beetles, bloodsuckers, caterpillars, cockchafers, eels, leeches, flies, grasshoppers, frogs, locusts, serpents, slugs, snails, termites, weevils and worms were disciplined by the ecclesiastical tribunals and in due time excommunicated.[5]

Because these smaller creatures couldn't be seized and imprisoned by the authorities, it was 'necessary to appeal to the intervention of the Church, and implore her to exercise her supernatural functions'. There is little evidence that these practices were ever particularly effective. Critics argued that this poor result might have been because the insects weren't baptised and members of the church in the first place. Priests argued that the poor outcomes were because the parish members hadn't tithed their meagre income to the church.

Chemicals for pest control

For thousands of years, stretching back to the ancient Sumerians, people have used chemicals to control pests. In the ancient Greek epic poem *The Odyssey*, written around 700 BCE, Homer mentions the use of 'pest-averting sulphur' – perhaps referring to the use of inorganic sulphur as a fumigant for lice. We also see the hero Odysseus burning sulphur to 'purge the hall and the house and the court'. Much later, in the 900s CE, arsenics were popular as pesticides, and from then onwards the use of metals as pesticides gained momentum. The highly toxic powder Paris green (copper acetoarsenite) was successfully used against the voracious Colorado potato beetle in the United States in the 1860s. It was a mainstay tool for the control of mosquitoes of the genus *Anopheles*, which spread malaria, and was widely used in many countries up until the mid 20th century.

The pesticide industry really began on a large scale with the arrival of DDT (Dichlorodiphenyltrichloroethane). The Nobel Prize for Medicine was awarded to the Swiss chemist Paul Müller in 1948 for the 'discovery of the high efficiency of DDT as a contact poison'. And DDT *was* highly efficient, at least initially, as an insecticide. But the indirect effects of DDT were substantial. They could be devastating for mammals, reptiles and birds, and many other groups of organisms. DDT drastically changed predator–prey interactions and even ecological communities. Science historian Hannah Gay describes just such a scenario with World Health Organization teams working in Sabah (previously North Borneo) and Sarawak, Malaysia. WHO teams had managed an indoor DDT spraying programme against mosquitoes during the 1950s.

Malaria mosquitoes were effectively controlled, but there were problems. For example, many people living in homes with thatched roofs complained that, after the spraying, their roofs were eaten away by caterpillars. It turned out that a parasitic wasp that preyed on the caterpillars was especially vulnerable to DDT, leaving the caterpillar populations less controlled. It was also the case that many domestic cats died. They were poisoned not only because they cleaned their fur by licking it, but because they caught and ate small geckos that lived inside people's houses. The geckos preyed on insects. The cat deaths appeared to provide early evidence of DDT working its way up the food chain, and that, at certain concentrations, DDT was harmful to mammals. The death of the wasps was a lesson in the loss of biological control. So, too, was the death of the cats, as the rodent population flourished. Replacement cats had to be trucked in to the most affected areas. In more remote areas, some were even parachuted in.[6]

A cat in a parachute would be a sight to behold.

When in 1962 the biologist Rachael Carson published *Silent Spring*, a book that alerted the world to the adverse effects of DDT on the environment, she was subject to a malicious and condescending rhetoric. An official with the Federal Pest Control Review Board drew laughter from his audience when he remarked, 'I thought she was a spinster. What's she so worried about genetics for?'[7] The director of the New Jersey Department of Agriculture wrote, 'In any large-scale pest program, we are immediately confronted with the objection of a vociferous, misinformed group of nature-balancing, organic gardening, bird-loving, unreasonable citizenry that has not been convinced of the important place of agricultural chemicals in our economy.' Other critics dismissed Carson as 'a priestess of nature' and 'a bird-lover', and even as a member of some mystical cult. (It's amazing how times have changed – many of those insults are now seen as positive qualities.) Today, Carson's *Silent Spring*, which presents a formidable set of statistics and a call to action, is widely considered one of the most important environmental books of the 20th century. Its arguments have inspired many environmental groups and government agencies.

DDT and a range of other chemicals were tested for social wasp control in New Zealand in the early 1950s. The powder was blown into a wasp nest via a tube connected to someone's mouth, or was scattered over the nest and entrance via a teaspoon borrowed from the kitchen. Occasionally, DDT would be applied to control aphids on oak trees, as a way of limiting honeydew collected by wasps.[8] It was a liberal time for pesticide use.

Other chemicals that were used include calcium cyanide and lead arsenate. These pesticides were often mixed with sugar to attract wasps, but this also attracted and killed insects like honey bees. Gammexane smoke generators were used, too. Gammexane, more commonly known as Lindane, is similar to DDT in that it has toxic effects on insects and, well, nearly everything else. The agricultural use of pesticides like Gammexane and DDT were banned in 2009 under the Stockholm Convention on Persistent Organic Pollutants.

We've learnt a lot since the days of DDT. But there are still questions. The most commonly used pesticides today are neonicotinoids, which are currently under restricted use in the European Union due to concerns over their effects on honey bees. There has been much debate around their use, and will be more to come. Many people believe that colony collapse disorder, a phenomenon when worker honey bees seem to abandon their queen, is largely due to the use of neonicotinoids, so they support restrictions or the complete ban of these pesticides. Neonicotinoids certainly can be lethal to insects like wasps and honey bees. There's also the question of how smaller concentrations of these pesticides may affect honey bees. Sub-lethal concentrations might make them susceptible to disease, or inefficient in their foraging, but these effects are very hard to measure.

Neonicotinoids clearly can affect bees, but what do they do in realistic conditions? I think it's very difficult to draw broad conclusions on the role of neonicotinoids. For example, a recent, 'largest-ever', multi-million dollar study was designed to determine the effects of neonicotinoids on honey bees in three countries. Statistically, this massive study found a mixture of beneficial and deleterious effects in Britain, only beneficial effects in Germany, and only negative effects in Hungary.[9] How should we react to these results? Despite even reputable sources such as the *New Scientist* reporting that this is one of two studies that provide the 'strongest evidence yet that neonicotinoids are killing bees',[10] I think it's difficult to make generalisations about these pesticides. And what happens if we do ban them? Opponents of neonicotinoid restrictions in the EU have argued that they are now forced to use older pesticides that have even worse effects on the environment and biodiversity. When the EU ban came into force, insects such as the flea beetle and peach potato aphid began to devour rapeseed crops in Britain.[11] Without more data, I'd be very reluctant to recommend that we ban neonicotinoids, causing growers to revert to spraying landscapes with older chemical pesticides that might have worse effects.

The debate on these and other pesticides rages on.

You need to kill the queen

Pesticides are the mainstay of social wasp control today. For many insects, the key to control is to kill a large number of the individuals on your crop or in your environment. But that's not so for wasps. To control wasps, you need to kill the queen. In nests of *Vespula* wasps, she is the only individual capable of producing more workers and new queens for next year. If the queen dies, effectively the nest dies, even though individual workers might live on for a few weeks. Spraying a single wasp worker with a can of fly spray might kill it, and perhaps a few of her nearby sisters, but typically there are thousands more wasps where that worker came from. And each of those workers is helping to raise even more. Killing a high proportion of the worker population might result in the collapse of the nest, but only if you are lucky.

Some sprays and powders contain chemicals like pyrethroids, for the purpose of regicide – queen killing. Both are sprayed or puffed into the nest entrance, usually at night. Powders cling to the wasps and are distributed throughout the colony as workers come and go. Pyrethroids work on the insect nervous system by altering the ability of signals to cross between nerve or axon endings. Historically, many pesticides have used a similar mode of action, such as carbaryl. These pesticides are neurotoxins – they're highly toxic to the nervous system of insects. They have considerably lower toxicity to vertebrates. Despite the low direct toxicity of carbaryl, there is evidence suggesting it is a human carcinogen. It certainly shows a high degree of toxicity to non-target organisms when sprayed into the environment. Consequently, older pesticides like carbaryl have been banned in many countries.

The best way to control wasps with pesticides is to tempt them with bait. That means you don't have to find every nest or climb every tree to deliver pesticide. With bait, a wasp can take a piece of the toxic offering, fly back to its nest, then share the food around its sisters and queen. The same wasp could make multiple trips to the bait and even lead others there. We know that wasps display social aggregation behaviour in their foraging – when a few wasps find food, wasps from other nests notice and go to investigate, so an extremely attractive food source can quickly accumulate large numbers of foragers from many different nests. Eventually, the lethal effects kick in.

There are a few caveats here. Firstly, you need bait that is highly attractive to wasps, and the wasps must be able to detect it from a distance. Secondly, you need a toxin that doesn't deter the wasps from taking the bait and that can be used at a low dose. It needs to kill the wasps, but not too quickly. With a low dose the wasps can make many foraging trips, so they take a lot of poison back to their nests. Thirdly, you need to have enough wasps around to find the bait and take it back to their nests. If the bait is used when wasp numbers are low or few foragers are flying, the chances of foragers from each nest finding the food and returning to the nest are low. Just the right numbers are needed.

The irony is not lost on biologists that many of us involved with conservation spend a lot of time killing things. Conservation managers, especially, would rather not spend their time this way. But, if they must kill pests, they want to use pesticides that don't harm other living things.

Ecologist Richard Toft recently developed such a bait and delivery system in New Zealand. He lives in Nelson, an area renowned for massive densities of wasps. As a boy Richard would watch his family's apple trees being eaten alive by wasps each year. Throughout his career of working with wasps, he's experimented with a range of toxins and baits for wasp control. Toxins tested before Richard came along included mirex, a pesticide widely used in the United States to control fire ants, but it was abandoned when it was found to be highly toxic to people and the environment. When Richard started working with wasps, chemicals like sodium fluoroacetate, better known as 1080, were being tested. But 1080 only worked on wasps when it was used in high concentrations. In the early 2000s, Richard and others began testing an insecticide called fipronil, and eventually developed it into a protein-based bait called Vespex. 'It only takes minuscule amounts of that toxin in a nest environment to actually have a lethal effect.' With the right bait, wasps bring fipronil back to the nest and feed it to other colony members. 'It takes a few hours so they've got plenty of time to circulate it. Once it's active it's very quick. The nest basically collapses overnight.'[12]

Biodiversity managers at New Zealand's Department of Conservation agree. 'When the wasp numbers get up high here the hum of the wasps in the trees is all you hear. We put the Vespex bait out and the next day you go into the forest and you can hear the birds singing, it's just amazing. It's like a switch,' says DOC biologist Nik Joice. A recent trial tested the efficacy of Vespex in an area of 5500 hectares of native beech tree forests. Wasp numbers fell by more than 93% and many more resources were available for native birds and other native biodiversity.[13]

One of the biggest hurdles with the use of fipronil and its development into a bait was caution around honey bees. No pesticide company wants to be associated with a product that might hurt honey bees. As Richard notes, 'There's absolutely no palatability commercially or environmentally for a product that is any threat to honey bees . . . the world's a very sensitive place without honey bees.'[14] Honey bees exposed to fipronil will die. The bait protein matrix in Vespex, however, is not at all

DOC biologist Nik Joice at a Vespex bait station. Worker wasps collect the bait, return to the nest and share it among larvae, workers and queen. *Photo: Dave Hansford*

attractive to honey bees, because it doesn't contain any sweet carbohydrates. The bait delivery system was also designed to limit harm on other organisms. As a result, Vespex can be used safely near beehives, and beekeepers use it the most.

The development and careful use of pesticides like Vespex as management tools is fantastic. Richard won a World Wildlife Fund award for Vespex in 2016.[15] However, he acknowledges that despite all the success, and even despite precautions and restrictions around its use, we still need to be mindful that it is a poison being introduced into the environment. He supports the idea of limiting the use of insecticides. 'I think the real issue with insecticides is not the product, it's the way they're used,' he says. 'I don't like them myself because I can see what potential damage they can have on the environment.'[16] The use and specificity of many pesticides is well advanced since the days of DDT. Most people, however, would prefer not to have to use pesticides at all.

A major problem is that many pests quickly develop resistance to pesticides. Such resistance frequently develops because the chemicals exert strong selection pressure on pest populations. In pest populations, there are often individuals that are different from the others. Initially, these individuals might be very rare. One might have, for example, a variation in the receptors of its nervous system. This variation results in differences in the way that signals cross between nerve or axon endings, and could have arisen because of a random genetic mutation. But what the difference means is that a neurotoxic pesticide like carbaryl doesn't work on the individual. These individuals are the only ones that survive when a farm manager sprays the neurotoxic pesticide. They survive, and breed. Repeated spraying of the pesticide repeatedly selects those individuals, and over many generations and applications of pesticide, the few become the many. The susceptible individuals die: they are weeded out from the population, leaving only the resistant.

That sort of resistance development is common. When resistance kicks in, we've seen new chemicals rolled out to replace the old ones. But new chemicals take time and money to find and develop, and fewer and fewer pesticides are being discovered.[17] There is a clear and urgent need to find alternatives. The ultimate goal for farm managers is to use a control method that is highly specific to the target pest.

Gene silencing

One such option might represent a major step change in pest combat. Gene-silencing, also known as gene knockdown and RNA interference, is a natural phenomenon that is emerging as a highly host-specific alternative to pesticides. It was discovered in 1998 by American biologists Andrew Fire and Craig Mello, who later won a Nobel Prize. Put simply, gene silencing is the regulation of gene expression in a cell to prevent the expression of a certain gene. The process of gene silencing occurs naturally in cells from a wide variety of organisms, from fungi to humans – in fact, you'll have some degree of gene silencing going on in your cells as you read this sentence. But scientists can also use this process selectively, meaning that gene silencing has many potential applications. To be clear, we're not talking about genetic modification in the traditional sense. Gene silencing doesn't occur within the nucleus of a cell; it doesn't alter or affect your DNA at all. It works via the 'middleman' of messenger RNA (mRNA).

Ribonucleic acid (RNA), along with DNA and proteins, is a nucleic acid that's essential for all known forms of life. When your cells want to make a protein that is encoded in your genes, they send a message to your cellular machinery using the 'middleman' mRNA. That mRNA is then translated into the protein. Gene silencing via RNA interferences works by destroying the mRNA and thus stopping the process of translation of genes on an organism's DNA into proteins or other cellular functions.

Gene silencing can be triggered by introducing foreign RNA into a cell. Let's say, for example, that this RNA is part of a virus. Your immune system has an enzyme called Dicer, which finds and then cuts that foreign, viral RNA into pieces. Those chopped-up pieces of viral RNA are then used by the cell to find and destroy any additional RNA that matches that specific sequence. The part of your immune system that enacts gene silencing is thought to have evolved specifically as a cellular defence against viruses. Gene silencing has now been co-opted to a variety of uses to stop or inhibit effective gene expression. We can create RNA artificially in the laboratory to uniquely target the exact sequence of a gene in a specific species. Provided you can get that RNA into cells, it can be designed to knock down or silence just about any gene. It can be targeted to insect skeletons, hormones, and digestive processes.

Gene silencing has been called the next generation of insecticides.[18] One group of scientists describe it as offering 'exquisite specificity and flexibility that cannot be matched by traditional chemical insecticides, biological control by natural enemies, or plants bearing protein-coding transgenes.'[19] Pest control methods using gene silencing are currently under development for a wide range of species – for instance, gene silencing for the blood-sucking parasite of honey bees, the mite Varroa, is expected to be commercially available by 2020. It has been successfully tested as a foliar spray to protect plants from herbivores.[20] It kills the plant-munching beetles, but leaves their ladybeetle-cousin predators untouched. Through gene silencing, biologists have also developed potatoes that express insect-specific RNA in their cells, resulting in the death of beetles that are feeding on the leaves.[21] And mosquitoes have been developed that have been modified to have gene silencing that inhibits infection by the Dengue virus.[22] Dengue fever infects between 50 million and 500 million people each year. If mosquitoes can't host Dengue then they can't transmit it, so this work has the potential to alleviate a major global disease burden. And, crucially for those of us concerned about wasps, gene silencing is in development for the control of them too, as scientists work on inhibiting a key part of their life cycle.

This approach to pest control isn't without critique. While gene-silencing technologies in sprayable formulations aren't modifying a genome, nor are they transgenic. Some still argue that gene silencing is like genetic modification and that we should therefore have major ethical concerns about it. The development of artificial genetic material, or RNA, also concerns some. One Australian group, the Safe Food Foundation, suggested that experimental wheat developed by the government would kill people.[23] They were worried that the RNA trigger being tested to change the wheat's starch content might match the gene for a human liver enzyme, meaning it would interfere with the liver too. In reality, however, that sort of problem isn't very likely at all. RNA sequences are typically very species-specific. You eat a huge variety of different RNA sequences in every meal without any ill effects. And when designing targets, scientists try to ensure specificity by comparing the sequences to massive DNA databases encompassing the genomes of many thousands of organisms.

The vast majority of scientists are confident of the specificity of approaches like gene silencing. But, as it's a new technology, we don't yet know every single effect it has. Most scientists would tell you, however, that it's a good thing to move away from broad-spectrum pesticides that are toxic to many organisms, and towards more specific, targeted methods. Though it requires considerable development and social acceptance before it can be used, I'm optimistic that gene-silencing or RNAi technology will offer a big step forward.

Tech controversy: Gene drives

Our knowledge of genomics is enabling another very different pest management approach – gene drives. Gene drives have been referred to as the 'fastest emerging tech controversy of the century'.[24]

Gene drives overcome the normal laws of inheritance (Mendelian genetics). With sexual reproduction, a trait can be lost or diluted through successive generations. For example, about one in 25 people are carriers of cystic fibrosis, a recessive disease. A carrier has two copies of the gene associated with cystic fibrosis, but only one of those gene copies contains a mutation; the other copy is normal. The person doesn't show any signs of cystic fibrosis, because only when both copies of the gene are fibrosis-positive will a person have the disease. However, they can pass their copy of the defective gene on to their children. If they have children with someone who doesn't carry the chromosomal defects, each child has a 50:50 chance of inheriting the genes for cystic fibrosis, because they inherit one of the two chromosomes containing the genes from that recessive parent. Consequently, there is a chance that none of the children by a recessive parent will be a carrier. That chance gets smaller and smaller with more children. If the person has three children, the chance of all three being recessive is $\frac{1}{2}$ x $\frac{1}{2}$ x $\frac{1}{2}$ = 0.125. There's a small chance that cystic fibrosis will be entirely lost from a family tree.

Any chance that your children won't inherit genes for a nasty disease is of course a good thing – but when we're thinking about invasive pests like wasps, and introducing genes that will help control them, that dilution is less ideal. Something called CRISPR-Cas9 could help solve this.

CRISPR-Cas9 is a gene-editing tool that scientists can use to remove, add or alter sections of the DNA sequence. They adapted this system from a similar process we see in some bacteria, where a gene-editing system has evolved to defend against invading pathogens like viruses. With CRISPR-Cas9, scientists can cut DNA at a very specific location and enable genes to be removed and/or new genes added. The amazing – and controversial – thing about this system is that once one chromosome has been edited, that modification is automatically transferred to the other chromosome. The changed gene, the guide to change the gene, and the machinery to do the change, are all automatically transferred. So, even if only one wasp has the CRISPR-Cas9 modification, all the chromosomes of all that wasp's offspring will carry that genetic editing. The trait is reliably passed down to the next generation and every generation thereafter.

Among insects, we've seen several natural cases of a type of gene drive. 'Selfish genes' or 'selfish chromosomes' have been known to invade and sweep through insect populations. For example, over a 60-year period one genetic trait in the fly *Drosophila melanogaster* copied and pasted itself within the fly genome and appears to have swept through all wild populations.[25]

Probably the number-one target proposed for gene drives is mosquitoes. Because of their ability to carry and spread diseases like malaria, mosquitoes are the deadliest animal in the world. (The second most deadly animals are human beings, killing other human beings, but that's another story.) Malaria causes between 500,000 and one million deaths per year, primarily in Africa. Dengue fever and the Zika virus are similarly widespread mosquito-borne diseases. So far, there are three broad approaches using gene drives to control the spread of such diseases. One is to target a gene that is essential to female fertility. A mosquito with only one copy of a defective fertility gene might be fertile, but any mosquito receiving defective genes from both parents would be sterile. Scientists in Britain, the US and Africa have been trying this, and the approach has proven successful in small laboratory trials. Another strategy is to distort the sex ratio of mosquito populations. It is possible to modify mosquitoes so that the X chromosome is selectively destroyed during sperm maturation. That means the resulting offspring would be non-biting males (it's only the female mosquitoes that transmit diseases). A third approach is to 'add cargo' to the mosquito genome – to insert genes that cause mosquitoes to become immune to the parasites that cause malaria, which belong to the genus *Plasmodium*. That approach has been met with some success in trials as well.[26] All of these approaches using gene drives will be developed for other pests. No one, as far as I'm aware, is currently developing gene drives for *Vespula* wasps. But gene drives are an option worth talking about.

One gene target suggested for a gene drive in New Zealand is a gene called *doublesex*, which is involved in determining the sex of insects. The expression of the gene could be manipulated by the CRISPR-Cas9 system so that no male wasps are produced. Without males, newly produced queens in autumn wouldn't be able to mate. The nests in the following year would fail. No workers or new queens would be produced. Scientists are considering other gene targets too, including those that lower reproductive rates or perhaps the rates of queens' survival through overwintering.

Any powerful technology holds risk and challenges, and with gene drives there are plenty of both. One such challenge is the genetic variation in the mosquito or pest population. The gene drive system relies on 'guide RNA', the RNA that informs the CRISPR-Cas9 system exactly where to make the insertion in the pest's genome. But if there is genetic variation within a species at the site of that guide RNA target, there can be a problem. It means the CRISPR-Cas9 gene drive will fail, and instead, over time, strains with that variation will become abundant and replace the strains that have been eradicated. It's very similar to what happens when pests develop resistance to pesticides. That's a major problem when targeting something like a particular mosquito species in Africa that spreads malaria. It's a massive area to target – it encompasses a huge range of genetic variation within the species of mosquito, as well as in the species of parasite that carries malaria.

Such substantial genetic variation is, however, much less of an issue with invasive species like wasps. Instead, it might actually be an asset, because there is little genetic variation in wasp populations in countries like New Zealand, which represents only a small fraction of the variation in the native range. For example, we currently estimate that there have been just two successful introductions of German wasps into New Zealand. By understanding the genetic variation here and in Europe, it would be possible to design gene drives with guide RNA that is specific to New Zealand populations and that would fail to have any effect on many of the wasp genotypes in their native range of Europe. There is a small chance that a genetically modified wasp queen could be inadvertently taken back to Europe from New Zealand, where it would be tragic to see a native species (even a wasp) eradicated. But a well-thought-out gene drive specific to the genotype of invaded populations would not lead to the obliteration of an entire species. American evolutionary biologist Kevin Esvelt, a pioneer of gene drive technology, refers to this type of approach as a precision gene drive. 'If a precision drive approach specific to invasives is feasible anywhere,' he told me, 'it would be your wasps.' Kevin is extremely cautious about using CRISPR technology for gene drives, however. 'The concern is not that it might spread into the native population and cause some kind of ecological impact. It's that it might spread back to the native population, period. That is, I am most concerned with social perception and public trust in science, as well as potential for backlash against the technology.' Other fail-safe options are being discussed and developed for gene drives, including 'daisy drives' that involve multiple genetic modifications. There are other challenges to implementing a gene drive, though I expect that over the next decade scientists will overcome many such challenges.

The potential of gene drives to for pest control is immense and extraordinarily powerful. As Jim Thompson in the *Guardian* argues:

> It's a perfect case of a very high-leverage technology. Archimedes famously said 'Give me a lever long enough and a fulcrum on which to place it, and I shall move the world.' Gene drive developers are in effect saying 'Give me a gene drive and an organism to put it in and I can wipe out species, alter ecosystems and cause large-scale modifications.'[27]

Even though this is invoking the classic 'mad scientist', to be fair, gene drive technology does seem to have that sort of potential. Kevin Esvelt has been widely quoted as saying 'Do you really have the right to run an experiment where if you screw up, it affects the whole world?'[28] He has a point.

To date, there have been no intended field releases nor unintended escapees. But the technology is spreading fast as scientists become increasingly aware of its potential. Many are urging extreme caution. In 2016, a position statement was released on the use of gene drives for conservation, jointly signed by the famous primate biologist Jane Goodall, the geneticist David Suzuki and many others:

We believe that a powerful and potentially dangerous technology such as gene drives, which has not been tested for unintended consequences nor fully evaluated for its ethical and social impacts, should not be promoted as a conservation tool . . . The invention of the CRISPR-CAS9 tool and its application to gene drives (also known as a 'mutagenic chain reaction') gives technicians the ability to intervene in evolution, to engineer the fate of an entire species, to dramatically modify ecosystems, and to unleash large-scale environmental changes, in ways never thought possible before. The assumption of such power is a moral and ethical threshold that must not be crossed without great restraint. . . . Given the obvious dangers of irretrievably releasing genocidal genes into the natural world, and the moral implications of taking such action, we call for a halt to all proposals for the use of gene drive technologies, but especially in conservation.[29]

Those are strong opinions, from a group of influential people.

In contrast, in the United States, the National Academies of Sciences, Engineering, and Medicine have concluded that we should take a moderate approach to gene drives and address people's concerns. In their 2016 report, they conclude that there is insufficient evidence to 'support the release of gene-drive modified organisms into the environment. However, the potential benefits of gene drives for basic and applied research are significant and justify proceeding with laboratory research and highly controlled field trials'.[30] They stress the importance of assessing the risks around issues ranging from gene-flow to unexpected trophic- and ecosystem-change. They also emphasise that when designing research programmes and making ecological risk assessments and policy decisions on gene drives, we need to factor in engagement with the public.

So, where should we stand on the use of gene drives? There are many diverse opinions and many more discussions to be had. But, as a scientist, I'm much more likely to pay attention to the well-researched and evidence-based document from the National Academies of Sciences, Engineering, and Medicine. The position statement by Jane Goodall, David Suzuki and others doesn't present anywhere near the type of evidence-based reasoning from which I'd expect governments to form policy and science direction. For this technology at the moment, I think the right call is 'proceed with caution'. Any use of gene drives for pest control is a few years away, but this technology is on the near horizon. It requires more data and more discussion. A recent review on the use of gene drives for wasp control in New Zealand concluded that:

> Perhaps the biggest barrier to adoption is the social one. For this technology to be effectively used in New Zealand, New Zealanders need to be well informed of the benefits and risks, and prepared to take those risks to live in a Vespine wasp-free environment.[31]

The old ways

An older but still effective approach for pest management is biological control, which uses a pest's natural enemy, such as a parasite or predator, to help control it. One of the earliest known examples of biological control is with the use of predatory ants by Chinese farmers. As far back as 1700 years ago, Chinese farmers used carnivorous ants to devour the insects that infested citrus trees. The farmers sold and bought ants for biological control at markets, and built bamboo bridges between the trees, so that the predatory ants could move from tree to tree.[32]

'Classical biological control' is when an invasive pest is introduced from an area and a natural enemy is then sought and introduced from the same native range. When classical biological control works, it can work spectacularly well. A textbook example is with the prickly pear cactus in Australia. These cacti were introduced into Australia as an attractive flowering plant, a natural agricultural fence, and in an attempt to initiate a plant dye industry. The first recorded introduction was in 1788. By 1920, prickly pear cacti covered 24 million hectares (58 million acres) of land, and they were spreading at a million hectares per year. The cacti invaded native areas and grazing land was rendered useless for farmers and settlers. The Australian government introduced more than a dozen insects to try to control the situation. One, *Cactoblastis cactorum*, the cactus moth, was sourced from its native range in Argentina, Paraguay and Uruguay. Its little orange and black striped larvae were brilliantly successful in chewing through millions of cactus hectares. By 1932, nearly seven million hectares – previously prickly, infested land – was clear, so was made available to 1165 settlers. Stagnating townships were revitalised, with public offices and entire communities rebuilt. A town hall in Boonarga, Queensland was named the Cactoblastis Memorial Hall, after the famous moth. Monuments were erected celebrating its consumption of cacti. Very few such memorials exist to honour insects anywhere. Both the prickly pear cactus and the cactus moth still exist in Australia, though at low densities. Nowadays most Australians would have little idea of how this little moth has benefited Australians and the Australian economy. In fact, that's why biological control can work so well: after a natural enemy is introduced and established, no other management is really required. Ideally, the pests and their control agents fade into the biodiversity background.

Several biological control options have been considered for wasps in New Zealand and other countries. There have been at least 150 insect species associated with wasps in their native range, some of which are predatory or parasitic. Of these, the parasitoid *Sphecophaga vesparum* was considered the best option for biological control; it was imported into New Zealand and released

The parasitoid *Sphecophaga* attacks the larvae of *Vespula* wasps. The adult female wasp crawls into a wasp nest and, if it is lucky enough to avoid the workers, lays its eggs on developing larvae and pre-pupae. When its eggs hatch, the larvae consume the wasp host alive. *Sphecophaga* is native to Europe and has been introduced to New Zealand in an attempt to exert biological control.
Photo: Bob Brown

in the 1980s. There are many, many parasitoid wasps like *Sphecophaga vesparum*. Some parasitise the eggs of other insects; others go for the larval, pupal or adult stages. *Sphecophaga* attack the older larvae or early pupal stages of wasps. The adult female has the unenviable task of crawling around inside a wasp nest, laying eggs in the cells. She must avoid lethal attack by wasp workers. When her eggs hatch, the parasitoid larvae lie in the cells, munching away on the living wasp larvae, which are eventually killed. The *Sphecophaga* pupates. And then its life history becomes really interesting.

The pupal *Sphecophaga* may emerge as: (a) an adult with no wings; this adult will spend its life in the nest, parasitising and potentially helping to kill entire nests; (b) an adult with wings, emerging after just a couple of weeks, then dispersing to find other wasp nests; or (c) a winged and hibernating form of the parasitoid, which overwinters in the nest, with both males and females being produced at this stage. Successful biological control of social wasps by *Sphecophaga* relies on the parasitoid form with no wings. That's because the wingless stage is the one that stays in the nest and effectively kills a high number of the workers.

By 1990, over 108,000 individual parasitoids were released in New Zealand. But the introduction of *Sphecophaga* has not resulted in successful control of wasps here. It has established itself, but it doesn't seem to have influenced wasp abundance. Only about one in five organisms released as biological control agents is successful, so perhaps this result isn't surprising. The failure of *Sphecophaga* may be related to limited genetic diversity, or perhaps a combination of a limited and 'wrong' genetic diversity. The previous releases here were essentially derived from a single nest, and thus potentially from a single female parasitoid. Such a limited genetic diversity might result in something called inbreeding depression, which basically means that insects mating with their sisters or other close relatives have weak offspring. In addition, that parasitoid was a genetic strain (*Sphecophaga vesparum vesparum*) sourced from Germany, Switzerland or Austria. Recent genetic analyses have indicated that both common and German wasps appear to have come to New Zealand, Argentina and other countries from England and western European countries like Belgium.[33] Thus, the strain of the parasitoid might not have co-evolved and be adapted with the specific strains of wasps in their invaded range. Those results suggest that it might be beneficial to source wasps from the now-known native range and import a high level of genetic diversity. Such a programme is underway in New Zealand.

Scientists are investigating other potential biological control agents of wasps. While it was once thought that social wasps don't suffer from parasites or disease, it is increasingly clear that they have a multitude of natural enemies, such as mites that feed on the sugary secretions of their larvae and appear to transmit a disease.[34] Such mites could be exploited to control their populations. And wasps share many diseases with honey bees (in one study, we observed every single wasp from several different colonies to be infected with viruses like the Kashmir bee virus).[35] There are parasitic flies that catch wasp workers while they are foraging. The fly paralyses the wasp and uses a can-opener-like ovipositor to lay an egg inside the wasp – then it buries the zombie wasp alive so that the developing parasitoid larvae can consume it at leisure. Horrible as this sounds, these natural enemies of wasps offer hope and potential for a self-sustaining, cheap biological control. Such optimism is to be encouraged. But there are also those who are more pessimistic. In a 1991 article for the *The New Zealand Journal of Zoology*, reviewing and critiquing the current research on wasps, Roger Akre concluded that 'there is no convincing evidence in the current research papers or in accounts published previously that biological control agents such as *Sphecophaga* will ever effect control of vespid wasp colonies.'[36] Such gloomy outlooks are still around today, and may well yet prove to be accurate.

Biological control is not without risk. Some of the biggest conservation problems in New Zealand are now from predators intentionally introduced for the classical biological control of invasive pests. Ferrets, stoats and weasels, for example, were introduced to control rabbits. Stoats are a major predator of native birds: they're one of three predators on a hit-list for complete eradication from New Zealand by 2050.[37] Attempts were made to introduce even more likely disasters: red fox and mongoose were brought to New Zealand, but failed to establish.[38] Despite such risks and a history of problems, personally I'm glad to see that people are investigating biological control for wasps. Biological control can be a self-sustaining and typically permanent method for pest suppression. When it works, it's very cost-efficient. And the same mechanisms that have led to social wasp success in countries like New Zealand (such as a reduced abundance of natural enemies) may mean that biological control agents are successful here too.

Other tools in the wasp control toolbox

Soon after the social wasp invasion into New Zealand, there were bounties on hibernating queen collections. In the spring of 1948, 3 pennies (the equivalent of about US$1 today) was offered for every German wasp queen sent to the Department of Agriculture. They received 118,000 queens (7000 of which were reportedly caught by one schoolboy). There was, however, no evidence of any reduction in the number of nests over the following summer.[39] On a personal note, I'd be enthusiastic for the government to reinstate such a programme now. While it almost certainly wouldn't influence wasp populations, I'm confident I could collect a million queens a year. That would result in a substantial reduction of my mortgage.

The use of pheromones, another option for pest control, is perhaps a more practical approach. Pheromones can be used to disrupt aspects of a wasp's life cycle. Mating disruption has been a mainstay in orchard pest control for decades, with pests like the codling moth, a type of fruit-eating moth. For such pests, individual dispensers of artificial female sex-pheromones are positioned in orchards. The pheromone dispensers are often placed on a card coated with glue. Attracted to the pheromone, male moths come to a sticky end, while the females go unmated and lay only infertile eggs. These sorts of approaches could be used to reduce the mating success of wasps too, if we were able to identify the right pheromone mix. But one issue with these pheromone- or pesticide-based control options is that they're useful only in the area where they're applied. Would a government pay for annual wasp bait or pheromones to cover millions of hectares of native forest every year? I'm guessing probably not.

Another option is the Trojan female technique, where typically the female passes on genes that would make male offspring infertile. Trojan female wasps would carry naturally occurring mutations that effectively limit their reproduction. To do this, we would need large facilities to rear Trojan wasps; we would then need to release millions of wasps carrying these mutations over large areas of land over a period of at least a decade, and possibly up to three decades. The mutant wasps would eventually come to dominate the population and perhaps even result in wasp extinction.[40] Scientists have had success using this approach with fruit flies in a laboratory. But do we really want massive wasp rearing facilities releasing

A researcher collecting wasp workers from a nest site in New Zealand. We also collected wasps in European countries to examine for differences in bacteria and viruses in wasps. Wasps in all countries carried a wide variety of pathogens, including the deformed wing virus. *Photo: Phil Lester*

millions of new wasp queens every year, for decades?

In New Zealand, as for many other countries, it seems likely that for now we are limited to potential options like biological control and gene drives for the control of pests over large tracts of land such as native forest. Perhaps tools like highly specific pesticides, gene silencing and pheromone use will be useful for wasp control on smaller scales. But we need to manage wasps over hundreds of thousands or millions of hectares. A review by Jacqueline Beggs concluded that 'estimates of ecological damage thresholds suggest that wasp density needs to be reduced by 80–90% to conserve the most vulnerable native species in honeydew beech forests. There are currently no control tools that will reduce density by this amount over large areas.'[41]

What management approach should we take for wasp control in these large areas? Every action and control method holds risk and consequence. Doing nothing is one option for management. But doing nothing would mean accepting the status quo – the local extinction of native plants and animals, the economic costs, and the annoyance and harm to people.

6. JUDGE, JURY, EXECUTIONER?

Conspiracy theories, scientific misconduct and a social licence to operate

In 2016, my university publicised our programme on wasp control as part of an advertising campaign to highlight its applied and hopefully useful research. There were great big wasp posters plastered around Wellington city and full-page wasp centrefolds in popular magazines. The giant wasp billboards must have been especially frightening for the entomophobes among us. (One estimate claims up to 6% of people suffer from entomophobia – a fear of insects – so the Godzilla-sized wasp probably wasn't welcome.) The university marketed the campaign to the public as a 'War on Wasps': 'Victoria scientists are devising novel ways to wipe wasps out – hopefully for good'. I liked the media programme and supported its promotion. It was a great way to engage and interest people. I typically interact with a limited audience of scientists, academics and students, so this advertising provided an unexpected bonus of gauging the perceptions of the public. My inbox received the full spectrum of responses.

The majority of comments were positive. Many people talked about their personal experiences with wasps. 'They destroyed my apples . . . I also suspect that they are responsible for the absence of silvereyes. I can also confirm the death of one of my chooks from wasp stings. If only the wasps could concentrate their hate on all the cats that invade my garden.' From another: 'Wasps are an uncontrollable menace, impacting on my family's lives and safety. I just want to send encouragement for your research into the eradication of this damn pest.' There were a few suggestions for novel methods for wasp management and control, several of which I'd not previously considered, nor attempted. 'I put a big sign up in the window, "NO Wasps", they blatantly ignored it, cheeky little gits!' Many shared their emotional responses. 'I can't even look at that thing without an overwhelming desire to stomp the life out of it.' The stated goal in our research programme is to dramatically lower numbers of wasps. Most of the comments I received were captured by this message: 'Lower their numbers. I'd feel happier if that number was 0. Hate them with a passion'.

An evil menace?

Then there was the other end of the spectrum of comments. There were a lot fewer of these responses, but they were equally valid perspectives. Andy Blick, an experienced conservationist, was the commenter who probably best

German wasp nest (grey, compared with the brown of the common wasp nest) found in autumn in Normanby, the North Island of New Zealand. The nest probably contained tens of thousands of wasps. They were poisoned prior to this couple being able to stand anywhere near it. In a mild climate queens can survive through winter, resulting in colonies becoming extremely large. *Photo: Fairfax Media NZ*

summarised the concerns some people have about efforts to control wasps. Andy told me that he had worked for the now defunct New Zealand Forest Service and then the Department of Conservation for many years. He is active on a range of environmental issues, including campaigning against the use of the mammalian toxin 1080 (sodium fluoroacetate) to control possums.

Andy emphatically stated that he wouldn't like to see New Zealand without wasps. He was correct in that the marketing campaign presented only one side of the wasp story – the negative side. 'I wanted to just raise your awareness that there are many other ways of looking at wasps,' Andy said, 'and that they are not the evil menace portrayed on your university's website.' In previous chapters I have acknowledged this perspective, and the fact that there are likely benefits to countries hosting wasps and other invasive species. Perhaps wasps help with pollination. Perhaps they aid in the control of garden pests that would otherwise gobble your cabbages. Perhaps, as a predator in New Zealand, they have even helped prevent the establishment of other invasive insects, such as the great white butterfly (*Pieris brassicae*). Without doubt, there are some people who appreciate and want to conserve social wasps, especially in their native ranges. Andy told me that he sees wasps as guardians, 'covering the forests in their protective cloak. Yes, they take, but they also give everything back, their lives and more.' As an example of the protection provided by wasps, he talked about their predation on the native insects that periodically outbreak and defoliate large areas of native forest in New Zealand. This is a natural and normal process. But the presence of wasps likely reduces these outbreaks, via their 'cloak of predation'.

'As scientists you should not be the decision-makers on these matters,' he argued. 'Your job is to present balanced research, not promote propaganda, or the vilification of any animal.' I can't think of a better way to capture Andy's full response to the advertising campaign than his statement: 'There is absolutely no balance to what's stated there, and as scientists, you should not be acting as judge, jury and executioner for any animal.'

To some, I'm sure Andy's comments are, well, just not true. We've heard the story of Janet Kelland being stung hundreds of times on her farm. She was extraordinarily lucky to live through the experience. After being covered in wasps she escaped only by frantically disrobing and throwing herself into a stream. I didn't ask her directly, but I'm more or less certain that Janet doesn't believe wasps provide a protective cloak. Similarly, parents who see their children regularly being stung won't support wasp conservation or protection. Nor will beekeepers who see wasps eating a third of their business, or conservation biologists who see native chicks being killed immediately after hatching. But Andy's comments will resonate with others. As we've seen, common wasps are a classic example of a biological invader – a species that has arrived with human assistance into a region where it is not native, then established itself and spread. Recently, however, the research field of invasion biology has become contentious among scientists, with some arguing that it should be reabsorbed into broader ecological fields, or disbanded altogether. Some have even labelled invasion biology a pseudoscience.

David Theodoropoulos writes in his book *Invasion Biology: Critique of a Pseudoscience* that 'contrary to the claims of invasion biologists, biotic assemblages are highly resilient, and readily accept and integrate newly-arriving species. These species in turn show many highly beneficial effects for their new communities, phenomena largely invisible to invasion biologists, or, when visible, are deliberately reinterpreted as "harm".'[1] He argues that there is no ecological rule nor reason why the distribution of any species should remain stationary. Everything moves around to some extent. Consequently, in his view, terms such as 'nativeness' and 'alienness' are meaningless. So too are terms such as 'invasion', 'ecosystem health', 'ecological harm', 'ecosystem integrity' and 'naturalness'. He firmly believes that there are no criteria by which we can measure native from non-native species; indeed, he argues that concepts of 'native' and 'non-native' species are illusory. More damning still is his argument that many of the phrases and terminology used by invasion biologists are xenophobic and militaristic. The media campaign calling for a 'war' against wasps is just such an example. Describing their work and their dedicated struggle to conserve native species, biodiversity managers and researchers describe themselves as 'locked in combat with unrelenting armies of exotics'.[2] In David's criticism of this terminology, he has a point. While many readers won't see militaristic, xenophobic or 'cancerous' undertones to these analogies, I'm sure others do.

David is blatantly wrong on several of his claims, including that there is no evidence that biological invasions are a major threat to endangered species.

Island nations like Guam and New Zealand have lost dozens of its native birds, with invasive predators a major contributing factor. There is absolutely no doubt that many bird species would still be with us today if it weren't for predators like snakes, rats, stoats and possums. And we see biodiversity flow-on effects in these communities, as plant communities change because they have lost their seed-dispersers. On Guam, for example, the brown tree snake invasion has been directly responsible for the extirpation of forest birds. In this case, 'the extirpation' largely equals 'the eating'. The reduced bird populations have in turn been responsible for a big decline in seed dispersal. The native plants and birds have co-evolved together, with birds consuming fruit and pooping seeds with a ready-mix of fertiliser. Because of their predation on birds and a dramatically altered ability for seeds to be dispersed, the brown tree snake appears to have caused a 61–92% decline in seedling recruitment.[3] The snake is changing the tree communities. This is just one of many examples of an invasive species dramatically altering entire island ecosystems.

David Theodoropoulos is perhaps at the extreme end of the alternative perspectives on biological invaders like wasps. He believes that invasion biology is deeply corrupted by money and that the field itself is a conspiracy theory. He believes that herbicide and pesticide industries have manipulated the field and are driving a false environmentalism, and that invasion biology is effectively scientific misconduct. Specifically, his concern is that scientists are too often highlighting the negative consequences of invasive species in order to obtain funding and resources to promote themselves and their work. I doubt that Andy Blick has read such publications, but effectively he echoed David when he wrote to me, 'In New Zealand natural sciences, I believe, people have become lost in the ideology, ego and money. They no longer look at things as they really are.'

I don't pretend to speak for all people involved in managing biological invasions. But I think that I can for many. I've seen first-hand their passion and dedication for the benefit solely of the native birds that they love. Their actions are not for personal gain or benefit. The weekends that small community groups sacrifice in order to set predator traps or poison wasps can be heard in birdsong. I can think of no better retort to David's conspiracy theory other than those notes directly from the beaks of these birds.

There probably are pesticide companies that have substantially benefited from invasive species management. And perhaps there are scientists whose careers have been advanced by the field of invasion biology. But I'm confident the majority are in this field because of their passion and dedication to biodiversity and conservation.

'Novel ecosystems'

We can find more moderate views in the mainstream science community. Mark Davis, a professor in terrestrial ecology and animal behaviour, published a paper co-authored with 18 other ecologists in 2011. Their paper, 'Don't Judge Species by their Origins', argues that the role of invasive species in biodiversity has been over-emphasised. 'Over the past few decades,' they write, '"non-native" species have been vilified for driving beloved "native" species to extinction and generally polluting "natural" environments. Intentionally or not, such characterizations have helped to create a pervasive bias against alien species.'[4] The authors acknowledge that, while there have been major extinction events on islands and in lakes, most 'invasive species' do not cause dramatic changes to communities. That's absolutely correct. In New Zealand, we have seen an array of invasive species become established. For example, most of the ant fauna here is exotic and fits the criteria for 'biological invader'. But almost none of these invasive species cause major harm.

Mark and his co-authors believe that biodiversity managers should embrace 'novel ecosystems', and incorporate most exotic species into management plans. They believe it is inappropriate to try to achieve the near-impossible goal of drastically reducing the abundance of exotic species or eradicating them altogether. Any management that does occur should be based on solid empirical evidence of 'harm'. Mark's personal view is that we have an ethical obligation for management only when an invasive species is causing harm to the health of people or the economy.

I talked to Mark about his work. 'If we have facilitated the introduction of a pathogen that directly threatens human health (whether directly through disease or undermining the food supply), then I think we would have an ethical obligation to the people affected by this pathogen,' he remarked. 'For introduced species that are just having ecological effects, I don't think we have an ethical obligation. We certainly may have preferences due to our particular value system, but I don't think we are ethically obligated to intervene. Basically, whoever has the power to make the decision will determine whether a response to an introduce species will be undertaken or not, again based on the value system of the individual(s) in power.' Mark felt that for those species not threatening human health or the economy, we should take a pragmatic approach, especially if public money is going to be spent. 'Of course, any private landowner can do what they want on their own property. When public money is involved, I think the managers/decision-makers need to work through the seven questions . . . before deciding to intervene.'

Mark's seven key questions focus on assessing the harm of the exotic species and of the cost and potential for effective management. Do the net effects on ecosystem services by this species truly constitute harm, or can the effects be regarded simply as change? Do cost-effective and environmentally acceptable management options exist? Can we learn to live with this invasive species?

Mark's article was met with what can only be described as horror by many in the biological invasion community. Daniel Simberloff, the author of *Invasive Species: What Everyone Needs to Know*, wrote a response in a subsequent issue of *Nature*, with the backing of 141 other ecologists. 'Pronouncing a newly introduced species as harmless can lead to bad decisions about its management,' he wrote. 'A species added to a plant community that has no evolutionary experience of that organism should be carefully watched. . . . The public must be vigilant of introductions and continue to support the many successful management efforts.'[5]

Personally, I don't entirely agree with much of the fuss over Mark's article. I don't think that he and his co-authors were saying that we shouldn't be concerned about species causing harm in that article. They specifically stated that biosecurity or quarantine efforts are worthwhile and should be encouraged. Their primary message was to urge biodiversity managers to prioritise which species to manage, based on an analysis of the species producing benefits or harm to biodiversity, human health, ecological services and economies.

Most of us would probably classify harmful species as 'pests'. Mark and his co-authors are saying that when we consider managing a species, the decision should be based on whether there is evidence that the species is harmful, not on where it has come from. Species that are more widespread and more harmful should be given a higher management priority. If an exotic species is found in only a small area, in small numbers with few effects, perhaps it should be given lower management priority. As we know, most of New Zealand's exotic ants would fall into that category. Many of these ants have come from Australia. They are established, they have been here for many decades, but they cause few problems. Why waste our time and money managing some Aussie ant when we have bigger problems? If we looked really hard, we could probably find some ecologically harmful effects that these exotic ants have. They probably speak in an Australian accent that is painful to the ears of other ants. But such effects pale into insignificance against a million hectares of the highest known densities of wasps in the world. The management of wasps would probably take priority over the management of Australian ants with odd accents. But we should also prioritise the spending of public money on wasps over other native and exotic pest species. If I'm correct that this argument is a valid takeaway from Mark's paper, then that seems fine to me.

But there are aspects of this stance that I don't find quite as easy to swallow. For example, there is clear evidence that the origin of many invasive species *does* actually matter. An Australian review study from 2007, for example, showed that the effects of exotic predators can differ from the effects of native predators that have evolved alongside their prey. We've seen that, typically, invasive predators have a much greater effect on their prey populations than do native predators.[6]

The effects of these introduced predators on prey populations were, on average, nearly double those of native and co-evolved predators. So, a single introduced predator has a greater effect than a predator that has co-evolved with its prey species. For example, the rock-wallaby in Australia has co-evolved with a range of predators, including the dingo. But an introduced fox now greatly limits, more than any other predator, the size and distribution of the rock-wallaby population in Australia. The effect-size of the red fox is nearly five times that of the rock-wallaby's native predators. The authors of this study suggested that their analysis

reveals an ongoing crisis because alien predators can have significantly greater suppressive impacts than native predators, keeping prey populations further from their predator-free population size, making them more vulnerable to stochastic [random] extinction forces, and thus leading to native biodiversity losses at regional scales.[7]

Many of us would also disagree with Mark's position on the responsibilities that people have with regard to invasive species. Ecologists like James Russell at the University of Auckland believe that we have an ethical obligation to manage invasive species for the benefit of our native species.[8] People brought predators such as stoats, weasels and wasps to New Zealand, and our fauna has suffered tremendously. I agree with James that we have an obligation to mitigate the further impacts of these exotic predators. And there is clear evidence that populations of native animals and plants from around the globe can benefit a great deal from the clearance of invasive predators.[9] But Mark's response to those of us who, like James Russell, feel an obligation to act, is:

Declaring harm is a value-based social decision, one that needs to be made through collaboration with the larger citizenry. This is not a scientific decision, even if scientists are making it. Russell states that the distinction between *native* and *nonnative* remains valid because 'our ethical duty to nonnative species … differs from our duty to native species.' Says who? Since its inception in the early 1980s, invasion biology has been rooted in a starkly value-based distinction between native and nonnative species.[10]

I suspect Mark's response would be supported by individuals who are anthropocentric or who are economically driven. Everyone has their own value system. I would hope that readers of a book like this have a passion for biodiversity and feel ethically obligated to prevent future extinctions. But what about everyone else?

Ranking the pests

If we accept that pest management is necessary, for ethical, anthropocentric and any other reasons, the next question is how to prioritise which pests to manage.

And how might our social wasps be ranked? A 2012 survey by James Russell on New Zealanders' attitudes to introduced wildlife and their control showed that attitudes to small herbivores such as possums and rabbits, and predators such as mustelids, rodents and wasps, were negative. (Participants were randomly chosen from the 2011 New Zealand national electoral roll, which limited the age of participants to a minimum of 18 years old, with a total of 801 people responding.) There was a strong consensus that all of these pests should be managed or even eradicated. Less than 1% of respondents thought that it was acceptable to do nothing.[11]

James's survey, and a 2001 study on public views about introduced wildlife,[12] suggest that people have a preference for ongoing, low-cost control operations over expensive one-off eradications. As James found, 'New Zealanders have more of a utilitarian (balancing the greater good) than protectionist (preservation for their own sake) attitude to introduced wildlife, but with important elements of sensitisation to global trends in biodiversity and conservation management, environmental pollution and animal welfare, stemming from various "new social movements"'.

The question of how to rank wasps was posed by asking the survey respondents to spend an imaginary $100 of tax money on introduced pest management. They had to decide how much to spend on the control of large herbivores (such as deer), possums, rabbits, wasps, and other predators or pests. Possums won the competition for control spending, with around a quarter of the money. But wasps were rated highly too, with $15. James's survey and a previous survey found that public attitudes were almost unanimous in categorising wasps as 'pests'. The rate of spending on wasps was surprising, given that the places where wasps are most problematic tend to be well away from the urban areas where most people live. Only a small minority of people perceived wasps as a 'resource'.

I think these surveys and studies provide an impetus or a licence for the government to spend money on the management of introduced wasps. A critique might be that such surveys don't incorporate a ranking for *all* important pests. But people who haven't studied our pests closely probably don't know enough to provide a sensible ranking. How might people rank common wasps against the exotic clover root weevil (*Sitona lepidus*) and the native porina moths (*Wiseana* species)? Both are pasture pests that have been economically damaging for farmers. I'm an entomologist and know a little about these species, but off the top of my head right now I couldn't provide a justifiable ranking of their pest status against each other or against common wasps. Similarly, what would a city-dweller in New York or Los Angeles know about the Russian wheat aphid (*Diuraphis noxia*) or Colorado potato beetle (*Leptinotarsa decemlineata*)? They might guess (incorrectly) where the pests originated and what type of insect they are (hopefully they'd get that part right). But would they have any idea how to rank them in a list of pests? Probably not.

We could try to develop a better way of ranking priorities for spending taxpayer money on pests or invasive species. This would mean analysing individual pest species. And I suspect that such attempts would be contentious and without clear conclusions. Most people would probably agree that a good pest-ranking system should incorporate a pest's degree of influence on human health, the economy, and on biodiversity and the environment. Some smart cookie could probably come up with a mathematical rubric that includes all of these effects to derive a single ranking number. However, this could lead to a lot of debate. Some people have strong anthropocentric views and value the economy above all else; others are motivated by and passionate about conservation; others believe that people have no ethical right to be judge, jury and executioner of any plant or animal. All of these positions probably hold some merit. Who should make the decision to manage or not to manage this or that particular pest? The current system in many countries is whichever industry or conservation group shouts the loudest or has the best lobbyist group and can get funding for the management of their favourite pests. The government often funds the loudest crowd.

Developing a pest ranking or priority system that encompasses multiple sectors is a challenge. But it is necessary if we want public money to be spent wisely. A key feature of these surveys was that there is no single 'public'. While we can see a consensus on some questions, there was also a divergent set of views. There were even differences between genders; for example, women want significantly more of their taxes spent on wasp control than do men. And attitudes change over time. For example, the use of poisons to control invasive pests has declined steeply over the last decade. There is a push now for control or extermination using multiple methods.

Indigenous peoples

In many countries, indigenous peoples hold an important role in the management of biodiversity. Aboriginal Australians, Native Americans and Māori in New Zealand all have claims and rights to the land, so these groups need to be part of the decision-making about biodiversity management – including decisions about biological invaders. In Northern Australia, for example, about 22% of land is owned by Aboriginal communities. Around 200 native title claims have been approved for 1.3 million square kilometres of land, nearly one fifth of the Australian continent. The Aboriginal sovereignty movement, which can be traced back to 1972, continues to push for the independence of Australia's 300-odd Aboriginal nations. 'When Captain Cook arrived here in 1770, he said he was claiming the continent in the name of the Crown. But on what legal grounds did he take our land?' is a well-justified perspective.[13]

New Zealand's founding document or agreement between indigenous Māori and the colonising Europeans is the Treaty of Waitangi, signed in 1840. For the management of biodiversity, the relevant sections are in Article 2 of the Treaty:

> Her Majesty the Queen of England confirms and guarantees to the Chiefs and Tribes of New Zealand and to the respective families and individuals thereof the full exclusive and undisturbed possession of their Lands and Estates Forests Fisheries and other properties which they may collectively or individually possess so long as it is their wish and desire to retain the same in their possession; but the Chiefs of the United Tribes and the individual Chiefs yield to Her Majesty the exclusive right of Preemption over such lands as the proprietors thereof may be disposed to alienate at such prices as may be agreed upon between the respective Proprietors and persons appointed by Her Majesty to treat with them in that behalf.

There has been and will continue to be ongoing debate around the Treaty, not least because the version written in te reo Māori holds different implications from the version written in English. The Māori version, Te Tiriti, effectively guarantees te tino rangatiratanga – the unqualified exercise of Māori chieftainship over their lands, villages, property and treasures. Māori signatories did agree to allow the Crown to buy their land, but it is not certain whether this translation sufficiently conveys the full implications.[14] Around 5% of New Zealand's land mass is currently owned and managed by iwi throughout the country. There are ongoing claims and court action for additional land and recompense for instances where the treaty was either ignored or misinterpreted.

I talked with Ocean Mercier about pest management and the history and current involvement of Māori in New Zealand's biodiversity. Ocean is an academic with Te Kawa a Māui, the School of Māori Studies, at Victoria University of Wellington. 'Māori also have their own stake in many industries that are affected by invasive species like wasps, such as the wine industry,' she said. 'There is real concern by some groups over their effects on industries and individuals.' But that concern and knowledge isn't universal. Ocean's students come from iwi all over the country, and I've given lectures to them on invasive wasps and pest control options. 'Most students just didn't know there was such a big problem with wasps – afterwards they were all fired up to go out and hunt them down,' Ocean said. 'Overall, there is a lack of attention being focused on pests like wasps, and a wider conversation is needed. There is a desire to be at the table, for Māori to be involved with biodiversity management at all levels, including for the management of species like wasps. But many Māori come from a background of historical concerns and distrust in governmental management. The Swamp Drainage Act of 1915 in New Zealand removed wetland ecosystems. The Public Works Acts that enabled urban development with biodiversity as a trade-off.' Ocean remarked that there is a feeling among Māori that their opinions and desires have been ignored. 'It's really about trust. A trust relationship between the people – the public – and those that make the decisions. If that relationship is healthy, then people are more willing to talk and work towards common solutions.'

I asked Ocean if there is room for gene drives and genetically modified organisms as an acceptable part of these solutions for Māori, and she noted that attitudes are changing. 'There will be a huge diversity of opinion in the Māori community regarding genetically modified organisms and gene drives. For many, this technology is becoming more acceptable. Diabetes, for example, is common and is treated using insulin from genetically modified bacteria.' But trust again becomes an issue. 'In the 1990s, New Zealand had a Royal Commission on the use of genetically modified organisms of pest control. The Māori submissions were diverse and encompassed "for" and "against" this technology.' In the end, it was not so much the overall outcome that was important to Māori; it was whether their voices had been heard. It was widely felt that Māori views were seen as 'cultural' and were weighed against 'science'. In the end, 'science' won and many Māori felt they had been bypassed. They were ignored. 'People want to feel like all options are on the table and that there is no secret agenda,' Ocean said.

A social licence to operate

Even when we have a lot of data and information, some people still mistrust science and scientists. As American surgeon and public health researcher Atul Gawande has written:

> the scientific orientation has proved immensely powerful. It has allowed us to nearly double our lifespan during the past century, to increase our global abundance, and to deepen our understanding of the nature of the universe. Yet scientific knowledge is not necessarily trusted. Partly, that's because it is incomplete. But even where the knowledge provided by science is overwhelming, people often resist it – sometimes outright deny it. Many people continue to believe, for instance, despite massive evidence to the contrary, that childhood vaccines cause autism (they do not); that people are safer owning a gun (they are not); that genetically modified crops are harmful (on balance, they have been beneficial); that climate change is not happening (it is).[15]

How can scientists change the minds of those who just don't trust science? Atul's view is that 'Rebutting bad science may not be effective, but asserting the true facts of good science is.' I think this advice is appropriate. In the future, when we think about new approaches to pest management, we need to assert the facts around the needs and risks.

Given the history of failed eradication programmes and biological control agents that have only made things worse, it's understandable why some people just don't want to place their trust in pest management. Stand-out failures include a countrywide attempt at red imported fire ant eradication in the

United States. From 1962–75, around 250,000 kilograms of mirex was spread by ground vehicles and aircraft. Heptachlor was also used. Many, many fire ants were killed, as were native ants. But not only did the fire ants persist – they rebounded in treated areas to even higher densities than before. Even worse, mirex biomagnifies through the food chain and is carcinogenic. The Stockholm Convention subsequently banned the production and use of mirex. It became known as one of a 'dirty dozen' of nasty chemicals.[16] History is full of these high-profile failed eradication attempts.

But seldom are the successful eradication campaigns given as much publicity as the unsuccessful ones. Red imported fire ants have become established in New Zealand on three separate occasions. And they have been eradicated on each occasion. These eradication programmes cost millions of dollars and involved spreading pesticide over large areas. But the success of these programmes and their value for New Zealand's biodiversity, health and economy, is immense. Pest management – and science as a whole – have improved since the days of mirex. As well as acknowledging the tangled history, we should openly celebrate our successes. Failures in pest control happen for both native and invasive pests, but success can have substantial, ongoing benefits that span centuries.

An open dialogue with our communities is clearly very important. Adaptive co-management, for example, is a collaborative governance model in which people share the responsibility and power for decision-making. There have been examples of co-management for invasive species, such as between Aboriginal communities and governments in Kakadu National Park in Australia's Northern Territory, looking at how to manage feral animals such as horses, pigs and water buffalo.[17] We also need tools such as social impact assessments that illustrate the human and community impacts of invasive species.

I think one potential way forward is offered in a document called *The Brussels Declaration: Ethics and Principles for Science and Society Policy-Making*. This document takes a grass-roots, evidence-based approach to policy-making. It involves politicians, science advisers, scientists (including social scientists), civil society leaders, clinicians and other academics with the aim of deepening people's understanding of how power operates in science and society, and how science can help us to make good decisions about the policies that affect all parts of our lives. Though it's still in development, *The Brussels Declaration* offers a blueprint for the ethics and principles to inform work at the boundaries of science, society and policy. It also attempts to explain why evidence plus dialogue doesn't often result in people making good decisions and laws, and that a different approach – a broader, multi-stakeholder and multidisciplinary one – is needed.[18] So while the declaration recognises that science is a fundamental pillar of society, it also recognises that interest groups have every right to have their voices heard.

Andy Blick echoed this sentiment. 'You need to carry the general public along with your thinking. You need to engage with those who disagree. You need an advocate for pests; an advocate for the wasps.' This approach will lead to a 'social licence to operate', where projects like wasp control are co-owned and there is political support and co-management of projects, advocacy, and a united front against critics. Where there is a high degree of confidence and trust that the people who manage invasive species will behave in a legitimate, transparent and accountable way.

To a large extent, the debate over whether to accept wasps and other invasive species as part of a 'new nature' or to control them is moot. A total of 196 countries around the globe are obligated to manage their invasive species. For instance, countries where common and German wasps have invaded, including Australia, Argentina, New Zealand and South Africa, are signatories to the 1992 Convention on Biological Diversity. One of the three objectives of the convention is to conserve biological diversity, and this obligates the signatories to 'prevent the introduction of, and control or eradicate, alien species which threaten ecosystems, habitats or species'. A revised and updated Strategic Plan for Biodiversity was produced from the tenth meeting of the Conference of the Parties, held in 2010, in the Aichi Prefecture of Japan. This plan specified 20 biodiversity targets. Target 12 states 'by 2020 the extinction of known threatened species has been prevented and their conservation status, particularly of those most in decline, has been improved and sustained'. Target 9 is 'by 2020, invasive alien species and pathways are identified and prioritized, priority species are controlled or eradicated, and measures are in place to manage pathways to prevent their introduction and establishment'. Consequently, there is a requirement for most of the world's countries to prioritise their invasive pests and manage them based on an evaluation of harm.

Andy Blick and others believe we should accept wasps as part of a new world in which we're living. Mark Davis and other scientists feel we have an ethical obligation to manage only the invasive species that harm us. But governmental signatories on the Convention on Biological Diversity are obligated to prioritise and manage biological invaders for the sake of biodiversity alone.

Together, we need to reach some sort of social licence to operate – to agree and gain public support on how to prioritise the invasive species and how to manage them – while recognising that the public will never have a single voice. There will always be some who disagree.

A wasp foraging for nectar on a clover flower.
Photo: Andrew Greaves

7. WHERE WASPS AREN'T DESPISED (QUITE SO MUCH)

Wasps in their native range

In 384 BCE, in a little city named Stagira in northern Greece, a man was born who would go on to make massive contributions to many diverse fields. He would make remarkably astute, accurate observations about botany, physics, linguistics, politics and government. He would tutor Alexander the Great. Obviously of most importance, though, and arguably his greatest contribution to learning, was his description of the biology of wasps and hornets in their native range. The observations contained in Aristotle's major work *Historia Animalium* ('The History of Animals') were simply without comparison for the next 2000 years. Aristotle had clearly studied the life cycles of wasps, hornets and many other animals. He described the initiation of a new nest. He described wasps' mating behaviour and how they build their nests. He knew that wasps lay eggs and knew how the larvae are fed. He described what happened at the end of the season. All of this knowledge can only have been gleaned from years of careful and detailed observation. Aristotle certainly wasn't perfect. He had some odd ideas about men in relation to women. He had odd notions that people who are less intellectually able should be enslaved. But wasps, he figured out pretty well.

However, over the next two millennia, the study of wasps and hornets took a sharp backwards turn. We entered the dark ages, big time. The extent of this growing entomological ignorance is perhaps best represented in the 15th-century compilation *Hortus Sanitatis*, widely considered the first natural history encyclopaedia. In it, we read that hornets 'dothe growe out of roten fowle horse flesche'. Hornets grow out of rotten, foul horse flesh? This display of ignorance and lack of observational power is astonishing. Though, the *Hortus Sanitatis* did get some aspects right, including how social wasps find and handle their prey. 'A waspe seketh her mete of stikinge carion, they have stinges like the scopio withinforth, and the(y) fetche there mete also frome the floures and frutes of the trees. They take flies and byte of their hedes and carie the(m) to their holes in the (e)rthe, but the moste parte of them leve by the caryo(n) fleshe'.[1]

'Uncontrollable terror'

Johan Christian Fabricius, born in Denmark in 1745, is widely considered one of the greatest entomologists of the 18th century. One of the 9776 insect species he named was *Vespula germanica*: the German wasp. That name would have seemed logical to Fabricius, as the particular specimens he examined came from Germany. These wasps are, in fact, widespread throughout Europe and have a native range extending from eastern Russia through to Britain. But, today, a German wasp in Britain can find itself faced with strange prejudice. 'Angry, drunk and unemployed German wasps are invading Essex' is a headline from a British newspaper in 2015.[2] An assumption these stories make is that the wasps, 'drunk' from feasting on fermenting fruit, are 'extra bold'. They also assert that the wasp workers have 'nothing to do' because they no longer need to bring food to the larvae in nests, as the wasp queens have stopped laying eggs. 'The wasps' sheer numbers, sudden freedom and potential drunkenness mean that they are now more likely to sting humans than usual.' In 2017 we heard, 'IT'S THE LUFTWASPE! German "SUPER" wasps who sting repeatedly threaten to invade British gardens and parks this summer'.[3] Dubious, attention-grabbing headlines aside, it's true that wasps appear strongly affected by feeding on fermented fruit. They can even become unable to fly and will fall over while walking. No one, however, is yet to prove the inebriated wasps are more aggressive. And they certainly aren't 'unemployed'.

It can be hard to find the facts when the media often prefer a sensationalist approach, but let's try. The common wasp, *Vespula vulgaris*, was named by Fabricius's Swedish supervisor and teacher, Carl Linnaeus, in 1758. Linnaeus named and classified only 3198 insects – but given that his speciality was botany, this is impressive. Today, the common wasp is extremely abundant in Britain. I'm not sure who took the time to count them, but one article puts the number at 240 billion.[4] The German wasp is less abundant. Their population numbers depend on favourable conditions such as mild winters and dry springs and summers. Nevertheless, it's the nasty 'jobless and drunk German wasps' that, we are told, will sting for no reason. The reporting that decisively tells us that German wasps cause all the problems is especially impressive, because you need some training to differentiate a common wasp from a German wasp. Most people who are stung wouldn't have a clue or care as to the exact identity of the culprit. Most folk also don't carry miniature breathalysers to check a wasp's sobriety. Nor do they question a villainous wasp on its employment status before attempting to flatten it with a rolled-up newspaper.

I wonder how *Vespula germanica* would be valued in Britain if Fabricius had named it *Vespula englandia* instead? Would there have been periods in which Germany, France or Ireland sought out and destroyed English wasps? Would the English wasps be reported as drunk and jobless? Perhaps the British people would have more national pride for their wasps' ability to eat pests.

Wasps and hornets have certainly been in the public eye for many thousands of years. Consider this verse from the Book of Exodus:

> I will send My terror ahead of you, and throw into confusion all the people among whom you come, and I will make all your enemies turn their backs to you. I will send hornets ahead of you so that they will drive out the Hivites, the Canaanites, and the Hittites before you.[5]

It seems that hornets – which are large wasps – were used as weapons to drive out entire nations. And this approach was successful, as Joshua reminds his people a few chapters later: 'Then I sent the hornet before you and it drove out the two kings of the Amorites from before you, but not by your sword or your bow.'[6]

As with nearly every bible verse, there's plenty of debate over whether this scenario is meant literally or figuratively. Some scholars suggest that the hornets were 'a figurative expression for uncontrollable terror'.[7] Or were the Jewish nation capitalising on a species that was expanding its range from one region into another? Many exotic species will attain huge densities after their initial incursion or expansion of their range, before their numbers decline to more ecologically sustainable levels. High densities of hornets might literally have induced enough terror to drive people from their homes. Sound a bit far-fetched? Waves of giant swarms and attacks by giant hornets (*Vespa mandarinia*, or yak-killers) are reported annually in China. Recently we've heard of people being chased for hundreds of metres by these hornets, and stung as many as 200 times.[8] Sounds like the sort of experience that would make me want to flee my home.

Around 2600 BCE, the Egyptian pharaoh King Menes used the Oriental hornet, *Vespa orientalis*, as a symbol for Lower Egypt. I'm not sure if he meant this as a compliment or an insult. Some authors have suggested that the hornet was probably intended to symbolise the fear spreading from the powerful monarch and new ruler. Regardless of his intent, he couldn't have chosen a more interesting hornet to represent this region. *Vespa orientalis* is characterised by yellow stripes across its otherwise bland and boring brown body. It has a yellow patch on its head as well. You might assume that the yellow stripes are

Vespa orientalis foraging for nectar. The yellow on the head and the thorax are derived from the pigment colour xanthopterin, which has a light-harvesting role, converting light into electrical energy. *Photo: imageBROKER / Alamy*

aposematic – a warning to predators or competitors. And the stripes probably do serve an aposematic function. But they also do much more. They function as a solar panel – a means to gather energy from the sun. This energy harvesting is a trick that, as recently as a decade ago, was thought to be limited to algae, cyanobacteria and plants. An Israeli research group have found that if you shine ultraviolet light on anesthetised hornets, they wake up faster, as though they have been recharged.[9] Additional experiments have revealed that the yellow surface has anti-reflective properties: light is absorbed into the yellow sections much better than over the rest of the body. The yellow stripes of *Vespa orientalis* are derived from the pigment xanthopterin, which has a light-harvesting role, converting light into electrical energy.*[10]

So, wasps in their native range have a long history of being maligned, misunderstood, and poorly reported. But they have also received much more careful, well-reasoned study.

Gangsters of the natural world

One of the strongest advocates for wasps in their native range is behavioural ecologist Seirian Sumner from University College London. Seirian, who has described herself as 'probably the biggest wasp-lover in the world', and who deems wasps 'the gangsters of the natural world', writes about the importance of wasps to the world's economy and to entire ecosystems.[11] In her article 'In Defence of Wasps: Why Squashing Them Comes with a Sting in the Tale', she talks about parasitic species of wasp. As we've seen, these solitary parasitoids lay their eggs in or on caterpillars or other insect hosts, and the eggs hatch and typically eat their host alive. Parasitoids are essential in keeping down numbers of pests around the world. Another species she discusses is the fig wasp and the role it plays as a specialist pollinator. 'The relationship between figs and fig

---

* The same research team who did these experiments capitalised on the widespread lack of animal ethics permission needed when undertaking research on insects. They discovered that should you chop off hornets' abdomens (and thus their solar panels), the hornets would 'flutter their wings when illuminated' but would be 'incapable of rising in the air'. Perhaps even less startling was the discovery that headless hornets do not fly at all.

wasps is arguably the most interdependent pollination symbiosis known to man,' she argues. 'Without one another, neither the fig nor fig wasp can complete their life cycle – a textbook example of co-evolution which is estimated to have been ongoing for at least 60 million years. Figs are keystone species in tropical regions worldwide – their fruit supports the diets of at least 1274 mammals and birds. The extinction of fig wasps would therefore be catastrophic in tropical ecosystems.'[12]

The pollinators that receive the most praise and celebration are, of course, bees. But social wasps can be important pollinators too, and, for Seirian, they are 'on a par with bees with the ecosystem services they provide.' Along with wasps, insects like flies, beetles, and butterflies are frequently overlooked as important pollinators.[13] We focus on honey bees, but we shouldn't ignore the whole insect community of pollinators. If species like bumble bees are lost from a community,

A young European paper wasp queen (*Polistes dominula*) on her newly built nest. This species looks very similar to the Asian paper wasp (*Polistes chinensis*) and was only noted here in New Zealand after an American entomologist collected wasps while visiting.
Photo: Art of Nature / Alamy

social wasps are able to take their place. In an experiment examining milkweed flowers, excluding bumble bees led to a near doubled rate of visitation by paper wasps in the genus *Polistes*. And those social *Polistes* were just as efficient at pollen transfer as the bumble bees.[14] Thus, despite the dire predictions we periodically hear, the world is unlikely to collapse if we lose bees from a particular area or region. Wasps and other pollinators just might be able to take their place.

Seirian also emphasises the vital role that social wasps play in pest control. She argues that without them, 'the planet would be pest-ridden to biblical proportions, with much reduced biodiversity. They are a natural asset of a world dominated by humans, providing us with free services that contribute to our economy, society and ecology.' She points out that a single nest of social wasps – with workers raising upwards of 10,000 larvae during the colony cycle – provides 'a whopping bang for buck in terms of ecosystem services, killing vast numbers of spiders, millipedes and crop-devouring insects.' The work in New Zealand that quantifies the massive prey intake by individual nests supports the assertion that wasps are major predators. While critics see their predation as one of the major problems of wasps, for Seirian it makes them extremely useful, 'minimising the need for toxic pesticides, but unlikely to threaten prey diversity. It is not yet possible to accurately quantify their huge economic value in this regard, but their diet of agricultural pests such as caterpillars, aphids and whiteflies makes a massive contribution to global food security.'[15]

Although it's unusual to hear of people enthusiastically defending social wasps, the beneficial role they play has been known for centuries. For farmers, a major advantage of wasps is fly control. In the 1800s, one observer counted the number of flies that were captured by wasps on just two cows. Between 300 and 400 flies were carried off in just 20 minutes. There are stories of English landowners who carefully destroyed all of their wasp nests, only to be plagued by flies. There is a story dating back to 1770 of an ingenious butcher who, each day, placed a calf liver outside his shop specifically to attract wasps; in turn the wasps defended his business against flies.[16]

In Asia, the yellow-legged hornet (*Vespa velutina*) feeds largely on flies. Observing this hornet preying on agricultural pests like biting stable flies and blowflies on the backs of pigs, one naturalist remarked in 1941 that this is 'a redeeming feature in an otherwise unattractive character'. Observations at that time show scientists paying close attention to such redeeming features. At one site in Indonesia, after a plague of caterpillars was eaten by the tachinid flies, Asian predatory wasps were observed to capture the flies. A wasp would hang by one leg from a thin branch of a shrub or tree, and, holding its prey with its other three legs, would gnaw on the fly.[17]

Wasps on farms

Scientists have been attempting to use social wasps for agricultural pest control for decades. Paper wasps, or social wasps in the genus *Polistes*, have been the chosen ones for this purpose. There are well over 300 different species of paper wasps, which get their name from their nesting habit of gathering wood fibres or plant stems to make their nests. Hanging from a sheltered location, a paper wasps' nest resembles a pinecone or a decomposing apple. A dozen or two of these inch-long wasps live in each nest, depending on the species and time of year. Paper wasps tend to be less aggressive than German and common wasps.

Polistes nests might look basic, and the wasps seem dully non-responsive when you wander up to their nests – but these are smart little animals. The stem of the nest, or petiole, is smeared with a chemical that repels ants from raiding the wasps' eggs and larvae. These wasps can also learn and remember the chemical signature of their nest mates, and, even more impressively, learn to recognise the faces of their sisters. The ability to recognise faces was once thought to be limited to animals with much larger brains. Researchers at the University of Michigan have demonstrated the learning ability of *Polistes* wasps by carrying out a novel face-painting experiment. These authors placed these paper wasps in small electrified mazes. The entire floor of the maze was electrified, with the exception of a 'safe zone' in one maze arm. To avoid getting a sub-lethal zap, the wasps had to learn to associate a small photograph of a wasp's face with that zone. There were two images to choose from: a wasp face that would result in a sub-lethal zap, and another random wasp face. The wasps clearly didn't appreciate being zapped and they quickly learned.[18] In their face-painting experiments, the researchers painted over facial markings to make known sisters now unrecognisable. Sisterly love did not result: instead, aggression flared because the wasps couldn't recognise each other or their social status within a nest.[19] More recently, our common vulgar wasps have also been shown to be able to recognise human faces. The study suggested that these insects can learn 'very complex but completely novel patterns like face images without any biological relevance'.[20] Armed now with this new revelation, next week I'll start training wasps to recognise and sting my more irritating colleagues within our department here at the university.

As far back as 1913, on the Islands of St. Kitts and St. Vincent in the West Indies, observers noticed that the Jack Spaniard wasp (*Polistes annularis*) helped to control the cotton worm: 'Planters are encouraging the wasps by erecting rough shelters for them in and near cotton fields.'[21] In the 1950s in North Carolina, nesting boxes were placed in tobacco plots in an attempt to increase *Polistes* nest abundance and pest predation. As described in a 1961 study, colonies of wasps were also successfully moved into experimental tobacco plots, which reduced hornworm populations by around 60% and overall damage to tobacco plants by

Nest boxes can be used in an attempt to increase natural populations of wasps for biological control of pests such as the cabbage white butterfly (*Pieris rapae*).
Photo: Jenny Jandt

74% – a great result. The wasps, however, had little or no effect on populations of tobacco budworms. The researchers suggested an integrated use of wasps with the pesticide TDE (tetrachlorodiphenylethane), an insecticide similar to DDT that was used on tobacco at the time.[22] Lucky smokers consequently got their tobacco sandwiched between two seriously cancerous chemicals.

Caterpillars feeding on plants have a repertoire of defences that aid their escape from predators, but employing these defences can cost them. The caterpillars will thrash their bodies around; they will regurgitate their plant lunch along with their acid digestive fluids; and, upon attack, they will try to bite the predator or drop from their host plant. Some caterpillars have spines that contain toxic chemicals which deter predators, although some wasps have learnt to gnaw these off to access the underlying meal of caterpillar flesh. Many caterpillars also have specialised behavioural defences. When harassed by wasps they will move away from the exposed (and better for foraging) part of the plant, where there are warm temperatures and new leaves, to more shaded parts of the plant. There, the plant leaves are of lesser quality and the temperatures are lower. So, the moths that develop from these harassed caterpillars are of poor fitness, due to relatively poor food and cooler temperatures. So, while many caterpillars will be eaten, the stressed and beleaguered individuals that manage to survive wasp jaws go on to produce fewer eggs. This sort of effect is a major sub-lethal impact of paper wasps. For example, the New England buck moth, an easily observable and gregarious moth that hangs out with its relatives, will hide deep within plant foliage to get away from wasps. One study estimated that of the effects of wasps on caterpillar survival, one third was due to the caterpillars fleeing to poorer quality parts of the plants.[23]

Other social wasps can have indirect effects on prey populations. For example, social wasps can help to spread viral pathogens that affect insect pests. The Swaine jack-pine sawfly, a major pest of pine trees in Eastern Canada, has been frustrating enough to farmers that red wood ants were introduced from Manitoba and Italy into Québec to help control it. The jack-pine sawfly is also vulnerable to a virus that can be sprayed onto trees from aircraft. Upon infection with the virus, the caterpillars become weak and lose their ability to defend themselves against predators. Decaying larvae won't be touched by the predatory ants; wasps, on the other hand, will eat just about any prey item, and they seem to home in on populations of virus-infected sawflies.

Writing in the 1950s, W.A. Smirnoff observed *Vespula* wasps catching diseased individuals, tearing off the abdomen, 'gluing' the anterior half to the needles, branches, or bark of jack pine trees, and flying off to the nest with the posterior of the insect. The wasps might return for the remaining carcass later. By gluing that carcass to the tree, however, the wasps were spreading the virus. Healthy larvae feeding on the tree then become infected. In 1959 this was a major discovery. 'To our knowledge no records have ever been made of predatory insects transmitting virus diseases to other insects,' wrote Smirnoff.[24] It's entirely possible that similar predatory activities of other wasps help spread diseases of other insects and pests.

Paper wasps have continued to be tested and encouraged for the biological control of pests in a wide range of crops, including cabbages – sometimes with some success. Predation by two native paper wasps in Brazilian coffee plantations seem to be an important factor in limiting outbreaks of the coffee leaf miner.[25] Other times, paper wasps aren't so effective, which results in high crop damage. It's often the case that these wasps act as effective biological control agents one year, but not the next. There are a variety of reasons for that, including that the wasps tend to forage on a range of crops. They will also attack many other insects, rather than focusing their energies on the pests on one specific crop. And, while social wasps eat major pests like the gypsy moth, outbreaks of such pests typically occur before social wasps have experienced their seasonal population build-up and the high densities that are ideal for pest control.[26]

Nevertheless, many scientists have concluded that paper wasps in their native range are worth encouraging as pest controllers, especially when the pest is abundant, large, easily seen and gregarious. And they are especially worthy of encouragement in places where there are few other pest control methods available, such as in developing countries. At least some degree of cheap pest control can be achieved by placing artificial nesting structures suitable for wasps near crops.

Wasps for pest control in New Zealand

Entomologist Barry Donovan has worked on wasps in New Zealand for four decades. In a 2003 paper he presents a detailed, well-considered proposal for the use of social wasps as generalist predators of insect pests, and writes that wasps have many characteristics that make them ideal for pest control. 'They have a large number of foragers per nest, a high demand for invertebrate protein, a very wide range of prey, an ability to tolerate low temperatures, a tolerance of nest manipulation, and a colony life cycle that can span at least two summers.'[27] We've already looked at wasps in beech forests and the high insect prey demands, and the resulting competition with native birds. When in agricultural environments, however, wasps compete with introduced birds such as sparrows. Most people aren't too concerned about an abundance of sparrows. (It's worth noting that, soon after the introduction of these birds to New Zealand in the mid 1800s, sparrows became so plentiful that they came to be regarded as pests, and 'Sparrow Clubs' were formed, wherein children would be armed with clubs and encouraged to club sparrows – but that's a gruesome story for another book.) Data from two nests of German wasps in New Zealand in agricultural environments highlight the voracious and hungry nature of the nests. The two nests were estimated to have collected 126 kilograms of prey.

This prey consumption was so high because the nests successfully overwintered, became very large, and survived for around 13 months.[28] A large portion of the wasp's food was flies. Honey bees and fragments of birds were found, though it was impossible to know if the wasps had scavenged them or attacked them while alive. It would be possible to facilitate this sort of successful wasp nest overwintering by placing nest boxes in protected environments. Such nest boxes have already been designed. Helping wasps overwinter might mean you'd see year-round biological control of invertebrate pests. Barry also suggests that using boxes would make the nests easier to manipulate, pointing out that the nests could be removed before fruits begin to ripen and become attractive to wasps, and before farm workers begin tending plants and harvesting fruit. Queen excluders, which are used in bee hives, could be used to prevent the escape of new queens and the spread of wasps. 'Farmers could have wasp nests available when and where required and in the numbers desired,' he writes.

'Wasps would be especially valuable for growers of some "organic" crops in which conventional pesticides are not permitted.'[29]

All of this sounds very sensible. Social *Vespula* wasps might well be efficient predators that could substantially reduce pest abundance. But there are some big issues to overcome with managing wasp nests. Wasps aren't bees. A lot of beekeepers, myself included, can wear minimal protection when working with our bees. Sometimes just a veil. In contrast, when my group and I are working with wasp nests, we have full reinforced bee suits, with duct-tape over the zippers and joints. Wasps seem to be able to locate the weak points of your suit and work hard to penetrate in more than one way. While the wasps from some nests are relatively non-aggressive, others seem to sting 'just because'. Part of the comparatively non-aggressive nature of bees is that they have undergone centuries of breeding programmes that select for less aggressive phenotypes and genotypes. Perhaps we could do the same breeding and selection for social wasps? There is also the serious issue that wasps are notorious raiders of beehives. The loss of beehives to wasps is a major economic cost that's driven by wasps in New Zealand.

Fifteen years after the publication of Barry's paper, he's still convinced that social wasps could be of benefit to pest control in New Zealand. 'In my experience wasps (and bees) don't attack if not interfered with, so appropriate education about their lack of propensity to attack if not disturbed should allay most people's fears, just as with honey bees.' Barry acknowledged that people would have to treat wasps with caution. 'Yes, wasps certainly can be very aggressive and in general much more aggressive than most honey bees, but this just means that appropriate protective gear would have to be worn. . . . Many situations require appropriate protection, including work with asbestos, many chemicals, deep sea diving, or space walking.' And, as he suggests, there is certainly a degree of hypocrisy in our views of wasps and bees. 'One sting can kill a person, and to that person all wasps are evil personified.' He points out that, in New Zealand, honey bees kill about one person a year and hospitalise hundreds. 'But because of the value of honey and pollination . . . there has never been a call for honey bees to be eliminated.'

Conservation in action

One of the best examples of social insects' importance to ecological communities in unexpected ways is the story of a butterfly known as the large blue (*Phengaris arion*) and a virus responsible for the rabbit disease myxomatosis in England. The large blue is a stunning and very collectable butterfly species. Many collectors would pay big money for specimens to add to their dead butterfly collection. In the 1950s, populations of living specimens were estimated at around 100,000 adults over a wide range of locations. But, just 20 years later, only one small population of the large blue was left. Why? People suggested it was the fault of over-zealous butterfly collectors, or air pollution, or even climate change.

One of the primary mechanisms for the butterfly's decline, however, was found to be the rabbit virus myxomatosis. As a young caterpillar, the large blue feeds peacefully on thyme. But as it turns into an older caterpillar, it develops a hunger for flesh – ant flesh. In its final stages of growth, the caterpillar develops special glands that secrete a sugary substance that seems to excite ants, and may even trick them into thinking that the caterpillar is a large ant larva. The ants are just so excited that they carry the caterpillar back to their nest. One species of ant is particularly important to the large blue because it is easily excited and fooled: the red ant, *Myrmica sabuleti*. Once the caterpillar has been carried to its new home, it spends the next 10 months chowing down on the young grubs of its ant host. Recently, scientists found that the caterpillar is even able to make sounds that mimic the ant queen. These sounds give the caterpillar a much higher social status among the ants.[30] The nest is safe as long as the ant colony obtains sufficient light and heat to stay alive. If the colony doesn't get sufficient heat, it will die – meaning that the large blue caterpillar will die too. That's where grazers like rabbits come into the story. With the introduction of the virus myxomatosis, rabbit populations declined, and the grass grew. With the grass free to grow tall, red ant nests became shaded from the sun, and died. No red ants means no large blue butterflies.

Once these interactions were illuminated, a conservation plan was successfully put in place.[31] My sons' favourite naturalist in the world, David Attenborough, commented: 'The restoration of the large blue butterfly to Britain is a remarkable success story, illustrating the power of ecological research to reverse damaging environmental changes.' He added, 'It is, moreover, a tribute to the dedication of many practical conservationists who have skilfully recreated its specialised habitat in our countryside.'[32] It is a fantastic, positive story of conservation in action. It's also a story of totally unexpected consequence of how our interference or management of one species (a rabbit virus) can distantly affect the survival and extinction probably of another species (butterflies) through the modification of food webs.

The story gets even more interesting when we learn that this genus of butterfly larvae, while living in ant nests, is attacked by parasitic wasps. The adult female wasp gains entry to the ant nest by releasing a chemical pheromone that causes the ants to attack each other. While the ants are all-consumed by angrily beating on their sisters for no good reason whatsoever, the wasps can attack the caterpillar in relative peace.[33] The authors of this study highlighted how this sort of discovery could be extremely useful as a potential new form of pest control 'The strength and persistence of this reaction suggests that similar cocktails of long-chain hydrocarbons might provide an alternative to the use of poisons and repellents to control pest ants'. Without any doubt, many species and awe-inspiring interactions, with potential benefits to people, are hiding right under our feet right now.

Eyes wide open

Ecological communities are complex. Given that a rabbit virus affects ants, butterflies and probably parasitic wasps, what unexpected consequences might await if we were to modify wasp abundance in the native or invaded range? Casual observations after nest removal in Britain suggest that we might expect more flies on our cows and in our kitchens. But in the invaded range, it's impossible to predict the outcome when we reduce numbers of a highly abundant invasive species. We can't expect that controlling or eradicating wasps won't affect animal and plant communities. Should we end up substantially reducing or even eradicating wasps from an invaded country like New Zealand, how would the loss of these generalist predators affect the agricultural and horticultural industries? We might end up with more flies. We might see more blowflies and maggots gnawing on New Zealand's many, many sheep (we have a lot of sheep). For many plants, fewer wasps might mean positive effects for their pollination, as a recent economic analysis on the effects of wasps concluded.[34] But for other plants, it could mean less pollination. We might see outbreaks of defoliating insects in native forests.

We should go into any future wasp management plan with our eyes wide open to these potential consequences, and expect outcomes for which we haven't prepared. But there would be many positive effects of wasp removal, probably in the primary industries, and definitely for birds like kākā and tūī. It is to those native ecosystems and our endangered species I believe we owe the priority of our efforts to manage invasive species.

So many of us seem to loathe wasps with a great passion. Wasps are greatly maligned even in their native range. But then there are scientists and conservationists who understand that wasps can play an important role in the functioning of ecosystems. Wherever they exist, wasps can be a nuisance; they can even injure or kill people, but they are an important predator and pollinator in some places. Whatever approach we take to manage wasps in their invaded ranges, we need to make sure that this management doesn't affect wasps, and the communities they live in, in their native range.

8. A WORLD WITH OR WITHOUT INVASIVE WASPS?

Deciding what the future holds

The Sunday night nature documentary this week is about Adélie penguins. It's impossible not to empathise with these little Antarctic birds as they eke out a living in harsh conditions. And it's easy to anthropomorphise them, just as the early, ill-fated explorers did in this part of the world: 'They are extraordinarily like children,' wrote polar explorer Apsley Cherry-Garrard, 'these little people of the Antarctic world, either like children, or like old men, full of their own importance and late for dinner, in their black tail-coats and white shirt-fronts – and rather portly withal.'[1] The narrator of the documentary doesn't mention the mating behaviour that explorer George Murray Levick described in 1914 as sexually deviant (it includes homosexuality, necrophilia and rape), an account deemed too shocking for publication at the time. The documentary instead focuses on a pair of penguins. The mother and father desperately try to keep their little pile of nesting stones from being stolen by their neighbours. While the father fusses over the family's stone pile, our heart rates surge as skuas approach and attempt to drag off and devour our little chick. Finally, having survived the skuas, the bland and infrequent invertebrate diet, and the freezing weather, the chick takes its first swim. Just as we're celebrating the little chick's success a heartless orca kills and eats it . . . on the juvenile penguin's first and now only swim. It had no opportunity to live its adult life, or to catch and eat krill that wasn't pre-digested.

The following Sunday night David Attenborough continues his Antarctic exploration. We learn how some populations of orca are comprised of matrilineal family groups, thought to be one of the most stable of any known animal species.

Orca have complex vocalisations and pass down their language and their hunting techniques to following generations. Sometimes they swim in unison to create a wave that washes tasty seals from their refuges of pack ice. These predators have a rudimentary animal culture, which, although contentious among scientists, is evidence of a high level of intellect; orcas can be placed in the same intellectual category as chimpanzees and human beings. We also learn that some orca populations are threatened or endangered due to prey depletion, pollution, habitat loss and conflicts with fisheries. We begin to hope that orca can catch and eat more of those penguins. There seem to be plenty of penguins, anyway, and besides, they have questionable moral standards.

You'll have figured out my point by now. It's fairly easy to present any species in a certain way. But no species is inherently good or bad. We tend to make those kinds of judgements based on a limited amount of information. It's the same with wasps. Our views are shaped by what little knowledge we have, and by personal experience (often the sharp, painful kind). And when wasps appear in the media, it's usually in a negative frame. We hear of stinging, interrupted picnics and sports days, anaphylactic shock, and deaths.

But I bet David Attenborough could make an hour of social wasp television that would pluck at your heartstrings.

This documentary would begin with a wasp queen waking up in spring. A cold, dark, snowy spring in Sweden. By some miracle, this queen – perhaps she is a solitary German wasp – survived the winter. She is the only one of her family alive. All 500 of her sisters have died. Some died because they chose an overwintering site that was too exposed to the rain and snow. Others died because they hadn't been fed enough by their workers during the summer, so they ran out of energy before spring and died of hunger in their sleep. Our single surviving queen is hungry and desperate to find a safe home in which to start her own family. Her first decision is whether to stay here in Sweden or make the long, arduous journey across the ocean to Denmark. If she stays, she and her young family might be found and consumed by voles. If she migrates, there is a chance she won't be able to survive the distance. She would drown, along with many other fleeing wasp refugees. Whatever she decides to do, once she establishes her small nest she finds herself fighting off other wasp queens that want to kill her and steal her nest. If she is lucky, she will encounter only one or two such challengers. Should our queen survive one, two or several potential usurpers, large predators like badgers will hungrily attack and try to eat her offspring. It's these predators that wasps need to protect themselves from. Wasps didn't develop their painful stings specifically with you in mind. Their weapons have evolved over millennia because of the desire of grumpy badgers and their acquaintances to eat entire wasp families. Should the queen survive these predators, her colony must overcome the plethora of diseases and parasites that threaten her home and her children. There is, in late autumn, a tiny, tiny chance that our queen has survived and will go on to produce new queens. The probability of her success is much less than 1%. But she makes it. The documentary cheers her cunning, skill, luck and success. She has beaten some seriously poor odds.

Social wasps like this German wasp queen offer much to be admired. They have an immense ability to learn. They can learn not just the faces of other wasps, but also how to interpret chemical cues from other species. Their sting delivers a tiny amount of venom that has evolved to cause maximum pain and long-term learning for the recipient. That minuscule amount of venom can even kill. The fact that so many of us loathe wasps can be considered a testament to their success.

Success! A worker wasp has caught a fly and is in the process of killing and dismembering it to take it back to her nest. Wasps have long been known to have a substantial appetite for flies. *Photo: Julien Grangier*

Even the invasive populations can be admired. The solitary queen of the common wasp managed to survive a journey halfway around the world to New Zealand in the 1970s. After her unexpected journey, she alone emerged from a shorter-than-usual hibernation period. The queen established populations where no common wasp had established them before. She found food and a home among a community full of strange new competitors and predators. She succeeded. People, too, once faced these challenges as they moved around the world, eventually occupying every continent. The wasp queen is here through no fault of her own: she was just looking for a safe place to shelter for the winter when her journey began. It was human beings who moved her from Europe and to a place where she took on a different status: that of a biological invader. She and her sisters bring benefits to their new homes, not least of which are their indirect effects on microbial communities, resulting in millions of tonnes of carbon sequestration. They help control outbreaks of flies and defoliating moths. They may play a role in the pollination of some plants.

But, as these wasps outside of their native range cause critical issues for many countries through their effects on biodiversity, on the economy, and on human health, by nearly any definition these wasps are pests.

Keeping perspective

When science writers talk about biological invasions, they often provide titbits of information about exotic species that support their argument for or against the management of that species. It's a little bit like tweeting – we have a limited number of characters to convey a message. A single tweet can't be anything but incomplete. So, an author will often use either the negative or positive effects of an invasive species to highlight some point in their current argument.

What is often portrayed, intentionally or not, is a biased perspective. For example, environmental journalist Fred Pearce discusses zebra mussels in his book *The New Wild: Why Invasive Species Will Be Nature's Salvation*. Zebra mussels are native to the freshwater lakes of southern Russia and the Ukraine. In the 1980s, these mussels were inadvertently introduced to the Great Lakes of North America – they were stuck to the hull of a ship travelling from the Caspian

Sea. The mussels spread widely in North America and to other countries, including the United Kingdom. But there appear to have been some benefits of this invasion for environments such as Lake Erie: according to Fred, the mussels turned out to be 'the best janitors Erie ever had'. Half a century ago, you could only see down for six inches – now you can see down for 30 feet in some areas. 'As light has penetrated the lake, some aquatic plants have revived. They in turn have become nurseries for fish such as the yellow perch.' A variety of species have been revived by the invasion, including previously endangered species such as the lake sturgeon, which likes to eat the mussels. A larger and encompassing meta-analysis of scientific studies in this system has demonstrated that some species benefit and others decline with invasions by zebra mussels.'[2] If you look hard enough, you will probably always find that a newly arrived invasive species brings some benefits, and in this case Fred seems interested mostly in highlighting the benefits of zebra mussels.

However, just as with wasps, there are two sides to the story. The cost of zebra mussels is substantial. They block pipes in waterways. They clog water intakes at hydroelectric companies. The cost of managing them runs into the hundreds of millions of dollars per year, and that's just for the Great Lakes of North America. The mussels might even be responsible for the outbreaks of deadly avian botulism poisoning that have occurred annually in Lake Ontario since their introduction. Prior to their arrival, this disease was not recorded for Lake Ontario. But since their arrival, the zebra mussel has become a major food source for another introduced species, the round goby, which is eaten by diving birds. As a group of Canadian scientists point out, the arrival of the round goby has probably 'provided a new efficient link for the transfer of benthic biomolecules, including Botulinum toxin, to fish-eating birds.'[3] An environmental journalist also argues that the mussels have effectively driven a 'deadly surge' in avian botulism, 'killing an estimated 80,000 birds, including loons, ducks, gulls, cormorants and endangered piping plovers' over the last 15 years.[4] In a review of *The New Wild*, Dan Simberloff pointed out that zebra mussels do bring some benefits, but they also threaten many native mussel species by smothering them and out-competing them for food.[5] These are examples of unexpected consequences of this biological invasion.

Should we even care about native mussels in the Mississippi? Botulism in birds? Should we worry about isolated species of butterflies and moths that are threatened by wasps in New Zealand's forests? Some people might. Many won't. But our governments and elected leaders have indicated, on our behalf, that we do and should. Nearly every country in the world has signed on to the Convention on Biological Diversity – conceived by the United Nations in 1988 – with the United States as one notable exception. As we saw earlier, one of the convention's targets (known as the Aichi Targets) is, by the year 2020, to prevent the extinction of known threatened species and to improve and sustain their conservation status, particularly those most in decline.[6] Other global organisations have similar goals, such as the United Nations Sustainable Development Solutions Network, which aims by 2020 to introduce measures to 'prevent the introduction and significantly reduce the impact of invasive alien species on land and water ecosystems, and control or eradicate the priority species'.[7]

Many have criticised the Aichi Targets and the global response to their implementation, with a large group of people stating that these goals are too limited and poorly defined. Most of the targets do seem to lack any explicit and quantifiable definition of success. One mid-term analysis indicated that 'despite accelerating policy and management responses to the biodiversity crisis, the impacts of these efforts are unlikely to be reflected in improved trends in the state of biodiversity by 2020'.[8] The trajectory for many of these efforts is, unfortunately, negative. The International Union for Conservation of Nature, comprised of government and civil society organisations, has called for even greater action towards protecting biodiversity and human wellbeing from invasive alien species.[9] To put it simply, we just aren't doing enough to combat what many call a 'biological annihilation' in the ongoing sixth mass extinction.

I'm also confident that many people will be of the view that we shouldn't have these targets or goals. Fred Pearce's opinion is that the Aichi Targets are 'pretty vague. But the presumption that there is something inherently "bad" about alien species is in my view wrong.' It seems to me that there is a continuum of stances on invasive species. Perhaps at one end are those who view all exotic species as 'bad' and believe that they should be managed or eradicated. Fred's stance might be at the other end. 'True environmentalists,' he writes, 'should rejoice when alien species burst through the paving stones of our cities, or wash up on foreign shores ready to colonise.'[10] But there must be limits to such a stance. If cholera were to once again seep through the paving stones of Broad Street in London, no one would welcome that biodiversity: the UK and a worried Europe would demand nothing less than eradication. If the virus causing foot and mouth disease were to appear, which could devastate a country's agricultural industry, the goal would again be to eradicate it. So there are limits to the exotic biodiversity that people are willing to accept. Not many people would welcome diseases or animals that feast on little children. Others might feel ambivalent about common wasps in New Zealand, even though they are an exotic species that, when in high abundance, are damaging our biodiversity and periodically killing people.

Should we really welcome everything, I asked Fred? He responded:

We are in the Anthropocene. We humans have a perfect right and arguable responsibility to manage ecosystems with that in mind. In particular, where species are doing us evident harm, using normal and appropriate (and fair) policy-making mechanisms, we should decide to cull or manage invaders whether diseases, pests or economic threats – or even sometimes for aesthetic reasons. My problem, in a nutshell, is our doing this in the name of somehow 'saving nature'. [His thesis is] to demonstrate how this is usually baloney, and based on some very misguided anthropocentric ideas about nature and how it works. Nature is a dynamic system, constantly changing, adapting, evolving. And invader species are an essential part of that.

Cherry pickers

A lot of invasion biologists are accused of cherry-picking the worst of the invasive species on which to focus their efforts. Fred Pearce sees these biologists being likened to the editors of tabloid newspapers: 'Like tabloid editors, they concentrate their studies on the nastiest and most sensational of invaders.'[11] But there is a failure of logic to this argument. Of course invasion biologists are going to focus their efforts on those pests or invasive species doing the most harm. Of course the government should use public funds to take action over the most damaging invaders and leave those that aren't harmful alone. And of course we shouldn't attack an exotic species just for being exotic if it does no harm. For these reasons I agree with Fred that invasion biologists 'seem to know little, and care even less, about either the silent majority of migrating species that just fit in or the great many places where there is little pandemonium.'[12]

I wonder if much of the debate around invasive species could be calmed if we instead referred to those species which do harm to our biodiversity and health simply as 'pests'. Sure, we would debate what classifies a pest. But it would become a less contentious debate, because 'pest' suggests a species that does harm, whereas an invasive species can establish itself in a new place and do little harm. There are many exotic wasps in New Zealand that don't appear to have caused extinctions or major biodiversity loss; nor are they detrimental to people or agriculture. They are typically solitary wasps that might buzz occasionally through your backyard and do little else. There is good evidence to believe that species from elsewhere are different from native ones, and they can potentially be pestiferous. But for the most part, I don't believe we need to focus our concerns on where this or that wasp species originally came from. For instance, even if a native wasp became so abundant that it is harmful to human health, economies and biodiversity, most of us would agree we should attempt to control it. And we know that social wasps over large parts of New Zealand, Australia, South Africa and South America cause us a lot of problems, so they easily fit the criteria for 'pest', even if we ignore their evolutionary origin.

Worryingly, this seems to be an area where some governments are ahead of us scientists. It's worrying because it is the hope of many scientists that science should inform policy, rather than the other way around. While we bicker over terminology, governments write legislation that avoids these issues. The New Zealand legislation that manages 'invasive species' is the Biosecurity Act 1993.[13]

Not once in this 300-page document will you find the word 'invasive'. Nor will you find the words 'alien', or 'exotic', or similar words that characterise invasive species. You will, however, find the word 'pest' used 320 times. I think that is showing some impressive foresight, though less impressive is that there is no actual information on how a pest is defined. The Act leaves the definition for regional councils to figure out, with the statement that 'pest means an organism specified as a pest in a pest management plan.' (Lawyers are great, aren't they?)

My local regional council's Regional Pest Management Strategy doesn't go on to define pests per se. It does talk about minimising 'the actual and potential adverse and unintended effects of pests on the environment, economy, biodiversity and the community'.[14] Both common and German wasps are included as pests in this plan based on their impacts on human health, beehives, viticulture, agriculture, forestry, and, of course, biodiversity. They are considered pests and thus, by law, require management.

Vespula vulgaris, the common wasp. *Photo: Colin McDiarmid*

The biodiversity your grandchildren will inherit

Over the last 20 years, we've become skilled in the use of toxins for pest control. In 2011, a rodent eradication project began on the remote island of South Georgia in the South Atlantic. It would be the largest rodent eradication attempt in history. Over five years, hundreds of tonnes of Brodifacoum were dropped as millions of pellets from helicopters. Now, for the first time in living memory since the eradication, rare birds such as the South Georgia pipit have returned and nested on the island. The director of the project, Tony Martin, said it is now very likely that the island is rat free, though this would require further monitoring. 'Already the South Georgia pipit and South Georgia pintails, both endemic species found only here, are returning in numbers we could never have imagined, along with other species which were the victims of rats.'[15]

Pest eradication projects like this will be further refined and enhanced in the years ahead. Science is rapidly advancing new technologies and techniques, and making massive leaps in fields like medicine. It was just a few decades ago that we discovered DNA; today, we can manipulate it. What will our new technologies enable us to do 50 years from now? Our ability to control harmful pests on large scales will continue to develop. For now, perhaps you and I should view our role as biodiversity caretakers for future generations, until new and safe technologies are ready to use. Techniques such as gene drives and gene silencing have the potential to replace the wide-scale use of chemical pesticides. Of all the approaches discussed in this book, I think there are major benefits to technologies like gene drives. They could be both humane and cost-effective. But there are ethical and technical considerations we need to make before we can safely use them. Other highly targeted approaches, such as biological control and gene silencing, hold immense possibilities too. Every approaach, even the approach of doing nothing and accepting this biodiversity as the 'new wild', comes with risk.

Whatever pest control method we use, I strongly believe that our techniques should be humane. In New Zealand there has been plenty of criticism over the use of chemical pesticides such as Brodifacoum and 1080, as these can lead to slow and painful deaths for pests, as well as for non-target animals. People have typically focused their thinking about the ethics of various methods on vertebrates, but we now know that invertebrates experience pain and stress too, as was clearly shown in recent experiments with crabs.[16] And we've found that common wasps don't appear to enjoy being given electric shocks either.[17] Insects do seem to have an ability to feel pain. They have nociceptors, which are sensory receptors that we know are central to the perception of pain in animals. Nociceptors allow insects to sense pressure and heat. They even allow insects to react to the chemicals which make peppers or mustard taste spicy hot.[18] Social insects like bumble bees, honey bees and wasps are certainly intelligent. They are able to learn and solve problems. Irrespective of how much people might loathe these little insects, given what we now know about them and how much there is still to learn, our wasp control methods should be as humane as possible.

Do *Vespula* wasps exert 'net harm' in their invaded range? I think the evidence is clear that they cause substantial economic loss, that they do affect human health, and that they do have a major impact on our biodiversity. Yes, they do appear to provide some degree of benefit, such as increasing carbon sequestration. They might help with plant pollination, or as a predator that aids in the elimination of potential invaders like the great white butterfly. But on balance, I think wasps do cause net harm in their invaded range. They are a fascinating and an ingenuous predator, but also a ruthless invader – a pest – that we have an obligation to manage.

It's my hope that my grandchildren will love and nurture our wildlife and biodiversity. I hope that science can provide them with safe and effective tools to manage wasps and other pests. And I hope that when our grandchildren walk through the beech forests of New Zealand, rather than the drone of wasps they will hear our wonderfully unique birdsong and smell an air sweet with honeydew.

What would you do about wasps?

A large common wasp nest that we discovered under the bark of this tree. Author Phil Lester (left) with Kevin Loope from the University of California.

Photo: Jenny Jandt

ACKNOWLEDGEMENTS

This book stands on the shoulders of many researchers and authors, without whom there would be little to say. There are a range of references that I've relied heavily upon, including Philip Spradbery's *Wasps: An Account of the Biology and Natural History of Solitary and Social Wasps*, Robin Edwards' *Social Wasps: Their Biology and Control*, Makoto Matsuura and Seiki Yamane's *Biology of the Vespine Wasps*, and Michael Archer's *Vespine Wasps of the World: Behaviour, Ecology and Taxonomy of the Vespinae*. All four are fantastic scholarly works. I've also drawn heavily on the research papers published in scientific journals over the last three decades. A substantial body of detailed and informative research has been developed in New Zealand. I'd like to particularly acknowledge the work and leadership of Jacqueline Beggs, Barry Donovan, Richard Harris, Henrik Moller, Joanna Rees and Richard Toft. I'd like to thank those people who took the time to talk with me, including Andy Blick, Mark Davis, Barry Donovan, Janet Kelland, Ocean Mercier, Charlotte Payne, Fred Pearce, Seirian Sumner, David Wardle and Geoff Watts. Many people read drafts and provided valuable feedback on chapters, including James Baty, Mariana Bulgarella, Peter Dearden, Eric Edwards, Mike Hannah, Sarah Lester and Richard Toft. An especially big thank you goes to Ashleigh Young, for her fantastic editing skills, interest and patience.

Work in New Zealand on these wasps has been supported by a range of government-funding agencies in New Zealand over many decades, including the Foundation for Research, Science and Technology, the Ministry of Business, Innovation and Employment, the Royal Society Te Apārangi Marsden Fund, and the Biological Heritage Programme under The National Science Challenge. I'm also grateful to Victoria University of Wellington and Victoria University Press for their support of this work.

NOTES

Epigraph: Stephen King, *The Shining* (1977; Anchor Books, 2013), 561.

Introduction

1 B. Hölldobler and E.O. Wilson, *The Leafcutter Ants: Civilization by Instinct* (New York, NY: W.W. Norton & Company, 2011), 89–106.

2 B. Hölldobler and E.O. Wilson, *The Ants* (Cambridge, MA: Belknap Press of Harvard University Press, 1990), 436–70.

3 D.W. Davidson, K.A. Salim and J. Billen, 'Histology of Structures Used in Territorial Combat by Borneo's "Exploding Ants"', *Acta Zoologica* 93 (2011): 487–91. doi.org/10.1111/j.1463-6395.2011.00523.x

4 J. Lubbock, *Ants, Bees and Wasps: A Record of Observations on the Habits of the Social Hymenoptera* (New York, NY: Appleton & Company, 1882), 1.

5 D. Mikkelson, 'Einstein on Bees', *Snopes*, accessed 27 October 2017. snopes.com/quotes/einstein/bees.asp

6 S.E. McGregor, *Insect Pollination of Cultivated Crop Plants* (Washington, D.C.: Agricultural Research Service, US Dept. of Agriculture, 1976), 1.

7 K.S. Delaplane, 'On Einstein, Bees, and Survival of the Human Race', UGA Honey Bee Programme, College of Environmental and Agricultural Sciences, accessed October 2017. caes2.caes.uga.edu/bees/

8 M.A. Aizen, L.A. Garibaldi, S.A. Cunningham and A.M. Klein, 'How Much Does Agriculture Depend On Pollinators? Lessons from Long-term Trends in Crop Production', *Annals of Botany* 103 (2009): 1579–588. doi.org/10.1093/aob/mcp076

9 B.I. Gherman et al., 'Pathogen-associated Self-medication Behavior in the Honeybee *Apis mellifera*', *Behavioral Ecology and Sociobiology* 68, no. 11 (2014): 1777–784. doi.org/10.1007/s00265-014-1786-8

10 Darwin, letter to Asa Gray, 22 May 1860, *Darwin Correspondence Project*, accessed October 2017. darwinproject.ac.uk/letter/DCP-LETT-2814.xml

11 D. Simberloff, *Invasive Species: What Everyone Needs To Know* (New York, NY: Oxford University Press, 2013), 2.

12 'Invasive Species', International Union for Conservation of Nature, accessed March 2017. iucn.org/theme/species/our-work/invasive-species

13 D. Carrington, 'Earth's Sixth Mass Extinction Event Already Underway, Scientists Warn', *Guardian*, 10 July 2017. theguardian.com/environment/2017/jul/10/earths-sixth-mass-extinction-event-already-underway-scientists-warn

14 G. Ceballos, P.R. Ehrlich and R. Dirzo, 'Biological Annihilation via the Ongoing Sixth Mass Extinction Signaled By Vertebrate Population Losses and Declines', *Proceedings of the National Academy of Sciences of the United States of America* 114, no. 30 (2017): E6089–E6096. doi.org/10.1073/pnas.1704949114

15 S. Lowe, M. Browne, S. Boudjelas and M. De Poorter, *100 of the World's Worst Invasive Alien Species: A Selection from the Global Invasive Species Database* (Auckland: The Invasive Species Specialist Group, a specialist group of the Species Survival Commission of the World Conservation Union, 2000).

16 C. Bellard, P. Cassey and T.M. Blackburn, 'Alien Species as a Driver of Recent Extinctions', *Biology Letters* 12 (2016): 20150623. doi.org/10.1098/rsbl.2015.0623

17 K. Thompson, *Where Do Camels Belong? The Story and Science of Invasive Species* (London: Profile Books, 2014), 121.

18 Ibid., 221.

19 H. Ledford and E. Callaway, '"Gene Drive" Mosquitoes Engineered to Fight Malaria', *Scientific American*, 30 October 2017. scientificamerican.com/article/gene-drive-mosquitoes-engineered-to-fight-malaria

1. When the babies feed the parents

1 P.J. Lester et al., 'Determining the Origin of Invasions and Demonstrating a Lack of Enemy Release from Microsporidian Pathogens in Common Wasps (*Vespula vulgaris*)', *Diversity and Distributions* 20, no. 8 (2014): 964–74. doi.org/10.1111/ddi.12223

2 R.J. Harris and J.R. Beggs, 'Variation in the Quality of *Vespula vulgaris* (L.) Queens (Hymenoptera: Vespidae) and its Significance in Wasp Population Dynamics', *New Zealand Journal of Zoology* 22, no. 2 (1995): 131–42. doi.org/10.1080/03014223.1995.9518030

3 J.G. Duman and J.L. Patterson, 'The Role of Ice Nucleators in the Frost Tolerance of Overwintering Queens of the Bald Faced Hornet', *Comparative Biochemistry and Physiology* 59, no. 1 (1978): 69–72. doi.org/10.1016/0300-9629(78)90308-0

4 M.E. Archer, 'Life and Fertility Tables for the Wasp Species *Vespula vulgaris* and *Dolichovespula sylvestris* (Hymenoptera: Vespidae) In England', *Entomologia Generalis* 9 (1984): 181–88. doi.org/10.1127/entom.gen/9/1984/181

5 See M.E. Archer, 'Population Dynamics of the Social Wasps *Vespula vulgaris* and *Vespula*

germanica in England', *Journal of Animal Ecology* 54 (1985): 473–85. DOI: 10.2307/4492. See also N.D. Barlow, J.R. Beggs and M.C. Barron, 'Dynamics of Common Wasps in New Zealand Beech Forests: A Model with Density Dependence and Weather', *Journal of Animal Ecology* 71 (2002): 663–71. doi.org/10.1046/j.1365-2656.2002.00630.x

6 P.J. Lester, J. Haywood, M.E. Archer and C.R. Shortall, 'The Long-term Population Dynamics of Common Wasps in their Native and Invaded Range', *Journal of Animal Ecology* 86, no. 2 (2017): 337–47. doi.org/10.1111/1365-2656.12622

7 E.L. Ormerod, *British Social Wasps* (London: Longmans, Green, Reader and Dyer, 1886; Ebook), 186. archive.org/details/britishsocialwa00firgoog

8 M. Masciocchi and J.C. Corley, 'Distribution, Dispersal and Spread of the Invasive Social Wasp (*Vespula germanica*) in Argentina', *Austral Ecology* 38, no. 2 (2013): 162–68. doi.org/10.1111/j.1442-9993.2012.02388.x

9 G. Rudebeck, 'On a Migratory Movement of Wasps, Mainly *Vespula rufa* (L.), at Falsterbo, Sweden', *Proceedings of the Royal Entomological Society of London* (A) 40 (1965): 1–8. doi.org/10.1111/j.1365-3032.1965.tb00292.x

10 K. Mikkola, 'Migration of Wasp and Bumble Bee Queens across the Gulf of Finland (Hymenoptera: Vespidae and Apidae)', *Notulae Entomologicae* 643 (1984): 125–28.

11 K. Vepsäläinen and R. Savolainen, 'Are Spring Mass Migrations of Bumblebees and Wasps Driven by Vole Cyclicity?', *Oikos* 91, no. 2 (2000): 401–04. doi.org/10.1034/j.1600-0706.2000.910221.x

12 R.W. Matthews and J.R. Matthews, 'War of the Yellow Jacket Queens', *Natural History* 88 (1979): 56–64.

13 H.C. Reed and R.D. Akre, 'Usurpation Behaviour of the Yellowjacket Social Parasite *Vespula austriaca* (Panzer) (Hymenoptera: Vespidae)', *The American Midland Naturalist* 110 (1983): 419–32.

14 P. Maschwitz, 'Das Speichelsekret der Wespenlarven und Seine Biologische Bedeutung'. *Zeitschrift für Vergleichende Physiologie* 53 (1966): 228–52.

15 J.P. Spradbery, *Wasps: An Account of the Biology and Natural History of Social and Solitary Wasps* (Seattle, WA.: University of Washington Press, 1973), 203.

16 C.L.R Payne, 'Perception and Practice of Entomophagy in Central Rural Japan', in *The Transactions of the Asiatic Society of Japan*, edited by C. Murray, R. Morton, K. McPhalen and E. Marx (Tokyo: Josiah University Press, 2015), 148.

17 C.L.R. Payne, *Liberty Ruth* (blog), 'Kushihara Hebo Wasp Festival'. libertyruth.com/blog/kushihara-hebo-wasp-festival

18 J. Mitsuhashi, 'Insects as Traditional Foods in Japan', *Ecology of Food and Nutrition* 36 (1997): 187–99. doi.org/10.1080/03670244.1997.9991514

19 C.L.R. Payne, 'Wild Harvesting Declines as Pesticides and Imports Rise: The Collection and Consumption of Insects in Contemporary Rural Japan', *Journal of Insects as Food and Feed* 1, no. 1 (2015): 57–65. doi.org/10.3920/JIFF2014.0004

20 R. Roberts, 'Hornets Taste Like Sausage', *The Food Geographer* (blog), 4 March 2015. thefoodgeographer.com/2015/03/04/edible-insects-hornets-taste-like-sausage/

21 R. Tackett, 'Alcohol Made with Fermented Wasps Gives New Meaning to the Phrase "Get Your Buzz On"', *Sora News 24* (blog), 6 April 2013. en.rocketnews24. com/2013/04/06/alcohol-made-with-fermented-wasps-gives-new-meaning-to-the-phrase-get-your-buzz-on/

22 Payne, 'Perception and Practice of Entomophagy in Central Rural Japan' (2015), 146.

23 Payne, 'Wild Harvesting Declines as Pesticides and Imports Rise' (2015), 60.

24 C.L.R. Payne and J.D. Evans, 'Nested Houses: Domestication Dynamics of Human–Wasp Relations in Contemporary Rural Japan', *Journal of Ethnobiology and Ethnomedicine* 13, no. 13 (2017): 9. doi.org/10.1186/s13002-017-0138-y

25 M.E. Archer, 'Successful and Unsuccessful Development of Colonies of *Vespula vulgaris* (Linn.) (Hymenoptera: Vespidae)', *Ecological Entomology* 6 (1981): 1–10. doi.org/10.1111/j.1365-2311.1981.tb00966.x

26 D.C. Schroeder and S.J. Martin, 'Deformed Wing Virus: The Main Suspect in Unexplained Honeybee Deaths Worldwide', *Virulence* 3 (2012): 589–98. doi.org/10.4161/viru.22219

27 J. Dobelmann et al., 'Fitness in Invasive Social Wasps: The Role of Variation in Viral Load, Immune Response, and Paternity in Predicting Nest Size and Reproductive Output', *Oikos* 126, no. 8 (2017): 1208–218. doi.org/10.5061/dryad.pb2rp

28 P.J. Lester et al., 'No Evidence of Enemy Release in Pathogen and Microbial Communities of Common Wasps (*Vespula vulgaris*) in their Native and Introduced Range', *PLoS One* 10, no. 3 (2015): e0121358. doi.org/10.1371/journal.pone.0121358

29 M.A.M. Gruber et al., 'Single-stranded RNA Viruses Infecting the Invasive Argentine Ant, *Linepithema humile*', *Scientific Reports* 7, no. 1 (2017): 3304. doi.org/10.1038/s41598-017-03508-z.

30 Q.-H. Fan, Z.-Q. Zhang, R. Brown and S. Bennett, 'New Zealand *Pneumolaelaps* Berlese (Acari: Laelapidae): Description of a New Species, Key to Species and Notes on Biology', *Systematic and Applied Acarology* 21 (2016): 119–38. doi.org/10.11158/saa.21.1.8

31 B.A. Klein et al., 'Sleep Deprivation Impairs Precision of Waggle Dance Signaling in Honey Bees', *Proceedings of the National Academy of Sciences of the United States of America* 107 (2010): 22705–2709. doi.org/10.1073/pnas.1009439108

32 E.P. Deleurance, 'Étude du Cycle Biologique du Couvain chez Polistes. Les Phases

"couvain normal" et "couvain abortif"', *Behaviour* 4 (1952): 104–15.

33 J. Dobelmann et al. (2017), 1208–218.

34 M.A.D. Goodisman, J.L. Kovacs and E.A. Hoffman, 'The Significance of Multiple Mating in the Social Wasp *Vespula maculifrons*', *Evolution* 61, no. 9 (2007): 2260–267. doi.org/10.1111/j.1558-5646.2007.00175.x

2. Spiders with no chance of survival

1 E. Arias E., M. Heron and J. Xu, 'United States Life Tables, 2012', *National Vital Statistics Reports* 65, no. 8 (2012): 1–68. cdc.gov/nchs/data/nvsr/nvsr65/nvsr65_08.pdf

2 J.R. Beggs and J.S. Rees, 'Restructuring of Lepidoptera Communities by Introduced *Vespula* Wasps in a New Zealand Beech Forest', *Oecologia* 119: (1999): 565–571. doi.org/10.1007/s004420050820

3 Ibid., 568.

4 R.J. Toft and J.S. Rees, 'Reducing Predation of Orb-web Spiders by Controlling Common Wasps (*Vespula vulgaris*) in a New Zealand Beech Forest', *Ecological Entomology* 23 (1998): 90–95. doi.org/10.1046/j.1365-2311.1998.00100.x

5 Ibid., 93.

6 R.J. Harris, 'Diet of the Wasps *Vespula vulgaris* and *V. germanica* in Honeydew Beech Forest of the South Island, New Zealand', *New Zealand Journal of Zoology* 18 (1991): 159–69. doi.org/10.1080/03014223.1991.10757963

7 Ibid., 159.

8 P. Rau and N. Rau, *Wasp Studies Afield* (Princeton, NJ: Princeton University Press, 1918), 295–96.

9 O.H. Wild, 'Wasps Destroying Young Birds', *British Birds* 20 (1927): 254.

10 J.P. Spradbery, *Wasps: An Account of the Biology and Natural History of Social and Solitary Wasps* (Seattle, WA: University of Washington Press, 1973), 144.

11 H. Moller, 'Wasps Kill Nestling Birds', *Notornis* 37, no. 1 (1990): 76–77.

12 Ibid., 76.

13 J. Grant, 'Hummingbirds Attacked by Wasps', *Can. Field Nat.* 73 (1959): 174.

14 J. Phipps, 'The Vampire Wasps of British Columbia: An Example of Haematophagy by *Vespula* sp. Bull', *Entomol. Soc. Can.* 6 (1974): 134.

15 Toft and Rees (1998), 94.

16 Beggs and Rees (1999), 568.

17 J. Beggs, 'The Ecological Consequences of Social Wasps (*Vespula* spp.) Invading an Ecosystem that Has an Abundant Carbohydrate Resource', *Biological Conservation* 99 (2001): 17–28.

18 R.J. Dungan, D. Kelly and M. Turnbull, 'Separating Host-Tree and Environmental Determinants of Honeydew Production by *Ultracoelostoma* Scale Insects in a *Nothofagus* Forest', *Ecological Entomology* 32 (2007): 338–48.

19 L.R. Crozier, 'Honeydew Resource Survey of Oxford State Forest', in *Papers Presented at the Honeydew Seminar* (Christchurch: Advisory Services Divison, Ministry of Agriculture and Fisheries, 1978), 15–25.

20 J.R. Beggs, B.J. Karl, D.A. Wardle and K.I. Bonner, 'Soluble Carbon Production by Honeydew Scale Insects in a New Zealand Beech Forest', *New Zealand Journal of Ecology* 29, no. 1 (2005): 105–15. newzealandecology.org/nzje/2252

21 Exodus 16:20 (New International Version): 'However, some of them paid no attention to Moses; they kept part of it until morning, but it was full of maggots and began to smell. So Moses was angry with them.'

22 H. Moller, J.A.V. Tilley, B.W. Thomas and P.D. Gaze, 'Effect of Introduced Social Wasps on the Standing Crop of Honeydew in New Zealand Beech Forests', *New Zealand Journal of Zoology* 18 (1991): 171–79. doi.org/10.1080/03014223.1991.10757964

23 J.R. Beggs and P.R. Wilson, 'The Kaka *Nestor meridionalis*, a New Zealand Parrot Endangered by Introduced Wasps and Mammals', *Biological Conservation* 56 (1991): 23–38.

24 G.P. Elliott, P.R. Wilson, R.H. Taylor and J.R. Beggs, 'Declines in Common, Widespread Native Birds in a Mature Temperate Forest', *Biological Conservation* 143, no. 9 (2010): 2119–126. doi.org/10.1016/j.biocon.2010.05.022

25 J. Carson, 'The "Alien Species" that's Making Our Native Birds Go Hungry – the Wasp', *Nelson Mail*, 9 January 2017. stuff.co.nz/nelson-mail/news/87271663/The-alien-species-thats-making-our-native-birds-go-hungry-the-wasp

26 Beggs et al. (2005), 111.

27 D.A. Wardle et al., 'Determining the Impact of Scale Insect Honeydew, and Invasive Wasps and Rodents, on the Decomposer Subsystem in a New Zealand Beech Forest', *Biological Invasions* 12, no. 8 (2010): 2624. doi.org.10.1007/s10530-009-9670-6

28 Ibid., 2632.

29 Ibid., 2634.

30 D.A. Wardle, P.J. Bellingham, T. Fukami and C.P.H. Mulder, 'Promotion of Ecosystem Carbon Sequestration by Invasive Predators', *Biology Letters* 3, no. 5 (2007): 479–482. doi.org/10.1098/rsbl.2007.0163

31 S. O'Donnell and S. Bulova, 'Development and Evolution of Brain Allometry in Wasps (Vespidae): Size, Ecology and Sociality', *Current Opinion in Insect Science* 22 (2017): 54–61. doi.org/ 10.101/j.cois.2017.05.014

32 Ibid., 54.

33 J. Hendrichs, B.I. Katsoyannos, V. Wornoaayporn and M.A. Hendrichs, 'Odour-Mediated Foraging by Yellowjacket Wasps (Hymenoptera: Vespidae): Predation on Leks of Pheromone-Calling Mediterranean Fruit Fly Males (Diptera: Tephritidae)', *Oecologia* 99 (1994): 88–94. doi.org/10.1007/BF00317087

34 M. Raveret Richter, 'Social Wasp (Hymenoptera: Vespidae) Foraging Behaviour', *Annual Review of Entomology* 45 (2000): 128. doi.org/10.1146/annurev.ento.45.1.121

35 T.S. Collett and M. Lehrer, 'Looking and Learning: A Spatial Pattern in the Orientation Flight of the Wasp *Vespula vulgaris*', *Proceedings of the Royal Society of London*, Series B 252 (1993): 129–34. doi.org/10.1098/rspb.1993.0056

36 Wits University, 'Dung Beetles Follow the Milky Way: Insects Found to Use Stars for Orientation', *Science Daily*, 24 January 2013. sciencedaily.com/ releases/2013/01/130124123203.htm

37 J.J. Foster et al., 'Stellar Performance: Mechanisms Underlying Milky Way Orientation in Dung Beetles', *Philosophical Transactions of the Royal Society* B 372 (2017): 20160079. doi.org/10.1098/rstb.2016.0079

38 A. Kelber et al., 'Hornets Can Fly at Night without Obvious Adaptations of Eyes and Ocelli', *PLoS ONE* 6 (2011): e21892. doi.org/10.1371/journal.pone.0021892

39 O.J. Loukola, C.J. Perry, L. Coscos and L. Chittka, 'Bumblebees Show Cognitive Flexibility by Improving on an Observed Complex Behavior', *Science* 355 (2017): 833–36. doi.org/10.1126/science.aag2360

40 Queen Mary University of London, 'Ball-rolling Bees Reveal Complex Learning', *Science Daily*, 23 February 2017. sciencedaily.com/releases/2017/02/170223142100.htm

41 Loukola et al. (2017), 833.

42 J. Grangier and P.J. Lester, 'A Novel Interference Behaviour: Invasive Wasps Remove Ants from Resources and Drop them from a Height', *Biology Letters* 7 (2011): 664–667. doi.org/10.1098/rsbl.2011.0165

43 Richter (2000), 129.

44 Ibid., 126.

45 *Bandits of the Beech Forest*, directed by Roy Hunt, written by Allan Baddock, produced by Michael Stedman (Natural History New Zealand and Shining Cuckoo Productions, 1996).

3. 'I put my foot right in the nest!'

1 C. Edwards, 'Wasp Attack Victim Cheats Death a Third Time', *Waikato Times*, 18 March 2014. stuff.co.nz/business/farming/sheep/9840622/Wasp-attack-victim-cheats-death-a-third-time. See also K. Bayer, 'Claims Woman Stung by Wasps Couldn't Get Through to 111', *NZ Herald*, 21 March 2014. nzherald.co.nz/nz/news/article.cfm?c_id=1&objectid=11224012

2 P.J. Lester et al., 'Determining the Origin of Invasions and Demonstrating a Lack of Enemy Release from Microsporidian Pathogens in Common Wasps (*Vespula vulgaris*)', *Diversity and Distributions* 20, no. 8 (2014): 964–74. doi.org/10.1111/ddi.12223

3 This area is renowned as 'wasp central' in New Zealand. My research group have observed 40 nests per hectare in the region. The biomass of these introduced wasps in this area can in some years exceed that of all the birds and introduced rodents and stoats combined. See C.D. Thomas, H. Moller, G.M. Plunkett and R.J. Harris, 'The Prevalence of Introduced *Vespula vulgaris* Wasps in a New Zealand Beech Forest Community', *New Zealand Journal of Ecology* 13 (1990): 63–72.

4 K. Bayer, 'Wasp Attack Victim Died of a Heart Attack – Coroner', *New Zealand Herald*, 6 February 2012. nzherald.co.nz/nz/news/article.cfm?c_id=1&objectid=10783779

5 D. Charpin, J. Birnbaum and D. Vervloet, 'Epidemiology of Hymenoptera Allergy', *Clinical & Experimental Allergy* 24, no. 11 (1994): 1010–15. doi.org/10.1111/j.1365-2222.1994.tb02736.x

6 P. MacIntyre and J. Hellstrom, *An Evaluation of the Costs of Pest Wasps (*Vespula *species) in New Zealand* (Wellington: Department of Conservation and Ministry for Primary Industries, 2015), 24. doc.govt.nz/Documents/conservation/threats-and-impacts/animal-pests/evaluation-pest-wasps-nz.pdf

7 R.E. Welton, D.J. Williams and D. Liew, 'Injury Trends from Envenoming in Australia, 2000–2013', *Internal Medicine Journal* 47, no. 2 (2017): 170–76. doi.org/10.1111/imj.13297

8 P.R. Notman and J.R. Beggs, 'Are Wasps More Likely to Sting Men than Women', *New Zealand Entomologist* 16, no. 1 (1993): 49–51. doi.org/10.1080/00779962.1993.9722650

9 P. Pečnerová et al., 'Genome-based Sexing Provides Clues about Behavior and Social Structure in the Woolly Mammoth, *Current Biology* 27 (2017): 2. doi.org/10.1016/j.cub.2017.09.064

10 N. St. Fleur, 'Male Mammoths Died In "Silly Ways" More Often Than Females, Study Finds', *New York Times*, 2 November 2017. nytimes.com/2017/11/02/science/mammoth-fossils-males.html

11 P.J. Lester et al., 'Critical Issues Facing New Zealand Entomology', *New Zealand*

Entomologist 37, no. 1 (2014): 1–13. doi.org/10.1080/00779962.2014.861789

12 See hortweek.com/scottish-natural-heritages-scottish-invasive-species-initiative-awarded-national-lottery-support/plant-health/article/1448040 and cbc.ca/news/technology/great-lakes-invasive-species-cost-u-s-200m-a-year-researchers-say-1.696402

13 See K. Thompson, *Where Do Camels Belong? The Story and Science of Invasive Species* (London: Profile Books, 2014).

14 D. Pimentel et al., 'Economic and Environmental Threats of Alien Plant, Animal, and Microbe Invasions', *Agriculture, Ecosystems and Environment* 84 (2001): 1–20. doi.org/10.1016/S0167-8809(00)00178-X

15 D. Pimentel, R. Zuniga and D. Morrison, 'Update on the Environmental and Economic Costs Associated with Alien-Invasive Species in the United States', *Ecological Economics* 52, no. 3 (2005): 273–88. doi.org/10.1016/j.ecolecon.2004.10.002

16 MacIntyre and Hellstrom (2015), 1.

17 'Wasp Scares Woman, Woman Jumps out of Car, Car Rolls into Sea', *Stuff.co.nz*, 6 January 2017. stuff.co.nz/motoring/88172990/Wasp-scares-woman-woman-jumps-out-of-car-car-rolls-into-sea

18 MacIntyre and Hellstrom (2015), 25.

19 Ibid., 22.

20 Ibid., 22–25.

21 B.K. Clapperton, P.A. Alspach, H. Moller and A.G. Matheson, 'The Impact of Common and German Wasps (Hymenoptera: Vespidae) on the New Zealand Beekeeping Industry', *New Zealand Journal of Zoology* 16, no. 3 (1989): 325–32. doi.org/10.1080/03014223.1989.10422897

22 MacIntyre and Hellstrom (2015), 8.

23 Ibid., 9.

24 Ibid., 8.

25 Ibid., 12.

26 Ministry for the Environment and Statistics NZ, *New Zealand's Environmental Reporting Series: Our Fresh Water 2017*. mfe.govt.nz/publications/environmental-reporting/our-fresh-water-2017

27 Ibid.

28 MacIntyre and Hellstrom (2015), 19.

29 R. Edwards, *Social Wasps: Their Biology and Control* (East Grinstead, UK: Rentokil Library, 1980), 215.

30 J.R. Beggs et al., 'Ecological Effects and Management of Invasive Alien Vespidae', *BioControl* 56 (2011): 505–26. doi.org/10.1007/s10526-011-9389-z

31 K. Haupt, 'Assessment of the Invasive German Wasp, *Vespula germanica*, in South Africa', Master's Thesis, University of Stellenbosch, South Africa, 2015.

32 Thompson (2014), 201.

33 MacIntyre and Hellstrom (2015), 35.

34 Ibid., 2.

35 Ibid., 34.

4. A cocktail of nastiness

1 M. Molloy, 'Woman Drives off a Bridge after 'Swerving to Avoid Wasp Inside Car'', *Telegraph*, 18 December 2015. telegraph.co.uk/news/newstopics/howaboutthat/12058043/Woman-drives-off-a-bridge-after-swerving-to-avoid-wasp-inside-car.html

2 'The Wasp Sting that Led to Involuntary Manslaughter', *Swiss Info*, 20 April 2017. swissinfo.ch/eng/nest-defence_the-wasp-sting-that-led-to-involuntary-manslaughter/43121810

3 J. Millsap, 'After Allergic Reaction to Wasp Sting Channahon Trustee Encourages All to Be Prepared', *Herald*, 22 August 2016, theherald-news.com/2016/08/21/after-allergic-reaction-to-wasp-sting-channahon-trustee-encourages-all-to-be-prepared/a11la8c/

4 D.B.K. Golden, 'Insect Sting Anaphylaxis', *Immunology and Allergy Clinics of North America* 27, no. 2 (2007): 261–72. doi.org/10.1016/j.iac.2007.03.008

5 M. Moreno and E. Giralt, 'Three Valuable Peptides from Bee and Wasp Venoms for Therapeutic and Biotechnological Use: Melittin, Apamin and Mastoparan', *Toxins* 7, no. 4 (2015): 1126–150. doi.org/10.3390/toxins7041126

6 A.M. Cará et al., 'The Role of Histamine in Human Penile Erection, *British Journal of Urology* 75 (1995): 220–24. doi.org/10.1111/j.1464-410X.1995.tb07315.x

7 T. Nakamura et al., 'Molecular Cloning and Characterization of a New Human Histamine Receptor, HH4R', *Biochemical and Biophysical Research Communications* 279 (2000): 615–20.

8 S.G. Cohen and M. Zelaya-Quesada, 'Portier, Richet, and the Discovery of Anaphylaxis: A Centennial', *Journal of Allergy and Clinical Immunology* 110, no. 2 (2002): 331–36.

9 Charles Richet, Nobel Lecture, 11 December 1913, accessed 27 November 2017. nobelprize.org/nobel_prizes/medicine/laureates/1913/richet-lecture.html

10 Ibid.

11 R.A. Wood et al., 'Anaphylaxis in America: The Prevalence and Characteristics of Anaphylaxis in the United States', *Journal of Clinical Immunology* 133 (2014): 461–67. doi.org/10.1016/j.jaci.2013.08.016

12 D.B.K. Golden (2007), 261.

13 M. Worm et al., 'First European Data from the Network of Severe Allergic Reactions (NORA)', *Allergy* 69 (2014): 1397–404.

14 P. MacIntyre and J. Hellstrom, *An Evaluation of the Costs of Pest Wasps (*Vespula *Species) in New Zealand* (Wellington: Department of Conservation and Ministry for Primary Industries, Wellington, 2015).

15 R.E. Welton, D.J. Williams and D. Liew, 'Injury Trends from Envenoming in Australia, 2000–2013', *Internal Medicine Journal* 47 (2017): 170–76. doi.org/10.1111/imj.13297

16 E. Barclay, 'Stinging Wasps Moving North Due to Warming?', *National Geographic News*, 16 July 2008. news.nationalgeographic.com/news/2008/07/080716-wasps-stings.html

17 R. Nuwer, 'When Animals Attack: Death Databases Indicate that our Fondest Phobias May Be Misdirected', *Scientific American*, 25 January 2011. blogs.scientificamerican.com/guest-blog/when-animals-attack-death-databases-indicate-that-our-fondest-phobias-may-be-misdirected/

18 Braun, 'Notes on Desensitisation of a Patient Hypersensitive to Bee Stings', *Bee World* 8, no. 10 (1927): 159. doi.org/10.1080/0005772X.1926.11096103

19 S.W. Ludman and R.J. Boyle, 'Stinging Insect Allergy: Current Perspectives on Venom Immunotherapy', *Journal of Asthma and Allergy* 8 (2015): 75–86. doi.org/10.2147/JAA.S62288

20 Z.-Q. Zhao et al., 'Descending Control of Itch Transmission by the Serotonergic System via 5-HT1A-facilitated GRP-GRPR Signaling', *Neuron* 84 (2014): 821–34. doi.org/10.1016/j.neuron.2014.10.003

21 J.O. Schmidt, *The Sting of the Wild: The Story of the Man Who Got Stung for Science* (Baltimore: John Hopkins Press, 2016), 230.

22 J.O. Schmidt, 'Venom and the Good Life in Tarantula Hawks (Hymenoptera: Pompilidae): How to Eat, Not be Eaten, and Live Long', *Journal of the Kansas Entomological Society* 77, no. 4 (2004): 405. doi.org/10.2317/E-39.1

23 Ibid.

24 Schmidt, *The Sting of the Wild* (2016), 100.

25 M.L. Smith, 'Honey Bee Sting Pain Index by Body Location', *PeerJ* 2 (2014): e338. doi.org/10.7717/peerj.338

26 Ibid., 4.

27 E. Yong, 'The Worst Places to Get Stung by a Bee: Nostril, Lip, Penis', *National Geographic*, 3 April 2014. phenomena.nationalgeographic.com/2014/04/03/the-worst-places-to-get-stung-by-a-bee-nostril-lip-penis/

28 'The 2015 Ig Nobel Prize Winners', *Improbable Research*, accessed October 2017. improbable.com/ig/winners/#ig2015

29 B. Grossi et al., 'Walking like Dinosaurs: Chickens with Artificial Tails Provide Clues about Non-Avian Theropod Locomotion', *PLOS ONE* 9, no. 1 (2014): e88458. doi.org/10.1371/journal.pone.0088458

30 L.A. Rosenberg, J.G. Glusman and F. J. Libersat, 'Octopamine Partially Restores Walking in Hypokinetic Cockroaches Stung by the Parasitoid Wasp *Ampulex compressa*', *Experimental Biology* (2007): 4411–417.

31 N.B. Leite et al., 'PE and PS Lipids Synergistically Enhance Membrane Poration by a Peptide with Anticancer Properties', *Biophysical Journal* 109, no. 5 (2015): 936–47. doi.org/10.1016/j.bpj.2015.07.033

32 Y. Kim et al., 'MP-V1 from the Venom of Social Wasp *Vespula vulgaris* is a de Novo Type of Mastoparan that Displays Superior Antimicrobial Activities', *Molecules* 21, no. 4 (2016): 512. doi.org10.3390/molecules21040512

5. *'We need to reduce their populations'*

1 L. Schmitz, 'Army Sent Flame Thrower to Destroy Huge Wasp Nest', *YouTube*, accessed October 2017. youtube.com/watch?v=wF-CdPFkZEo

2 Charlie Mitchell and Sam Sherwood, 'Wannabe Wasp Killer Told 'Leave It to the Professionals' after House Fire', *Stuff.co.nz*, 22 April 2016. stuff.co.nz/national/79208386/man-tries-to-kill-bugs-accidentally-sets-fire-to-house

3 K. Grant, 'Man Sets Fire to Neighbour's House Trying to Smoke out Wasp Nest', *Telegraph*, 4 September 2013. telegraph.co.uk/news/picturegalleries/howaboutthat/10285492/Man-sets-fire-to-neighbours-house-trying-to-smoke-out-wasp-nest.html

4 A. Ernst, 'Spontaneous Combustion of Wasps' Nests', *Nature* 18, no. 1 (1878): 487–88. doi.org/10.1038/018487e0

5 H.L. Carson, 'The Trial of Animals and Insects: A Little Known Chapter of Mediæval

Jurisprudence', *Proceedings of the American Philosophical Society* 56, no. 5 (1917): 413. jstor.org/stable/984029

6 H. Gay, 'Before and After *Silent Spring*: From Chemical Pesticides to Biological Control and Integrated Pest Management – Britain, 1945–1980', *Ambix* 59, no. 2 (2012): 97. doi.org/10.1179/174582312X13345259995930

7 F. Graham, 'Fifty Years after *Silent Spring*, Attacks on Science Continue', *YaleEnvironment360*, 21 June 2012. e360.yale.edu/features/fifty_years_after_rachel_carsons_silent_spring_assacult_on_science_continues

8 C.R. Thomas, *The European Wasp (Vespula germanica Fab.) in New Zealand* (Wellington: Department of Scientific and Industrial Research, 1960), 65.

9 B.A. Woodcock et al., 'Country-specific Effects of Neonicotinoid Pesticides on Honey Bees and Wild Bees', *Science* 356, no. 6345 (2017): 1393–395. doi.org/10.1126/science.aaa1190

10 S. Wong, 'Strongest Evidence Yet that Neonicotinoids are Killing Bees', *New Scientist*, 29 June 2017. newscientist.com/article/2139197-strongest-evidence-yet-that-neonicotinoids-are-killing-bees/

11 L. Gray, 'Neonicotinoid Ban Hit UK Farmers Hard: Bugs Devour Rapeseed Crop in Britain as EU Ban on Pesticide to Save Bees Comes into Force', *Guardian*, 1 October 2014. theguardian.com/environment/2014/oct/01/neonicotinoid-uk-farmers-rapeseed-crop-bees-pesticide

12 J. Carson, 'The Weapon to Wipe out Wasps: The Story of Vespex | Wasp Wipeout', *Nelson Mail*, 17 January 2017. stuff.co.nz/nelson-mail/news/wasp-wipeout/87865462/the-weapon-to-wipe-out-wasps-the-story-of-vespex--wasp-wipeout

13 E. Edwards, R. Toft, N. Joice and I. Westbrooke, 'The Efficacy of Vespex® Wasp Bait to Control *Vespula* Species (Hymenoptera: Vespidae) in New Zealand', *International Journal of Pest Management* 63, no. 3 (2017): 266–72. doi.org/10.1080/09670874.2017.1308581

14 As quoted in Carson (2017).

15 'NZ Innovation Shines in War Against Wasps', *WWF,* accessed September 2017. wwf.org.nz/?14201/NZ-innovation-shines-in-war-against-wasps

16 As quoted in Carson (2017).

17 B. Borel, 'CRISPR, Microbes and more are Joining the War Against Crop Killers', *Nature* 543, no. 1 (2017): 302–04. doi.org/10.1038/543302a

18 K.S. Miguel and J.G. Scott, 'The Next Generation of Insecticides: dsRNA Is Stable as a Foliar-Applied Insecticide', *Pest Management Science* 72, no. 1 (2015): 801–09. doi.org/10.1002/ps.4056

19 J.G. Scott et al., 'Towards the Elements of Successful Insect RNAi', *Journal of Insect*

Physiology 59, no.12 (2013): 1212–221. doi.org/10.1016/j.jinsphys.2013.08.014

20 Miguel and Scott (2015), 807–08.

21 J. Zhang et al., 'Full Crop Protection from an Insect Pest by Expression of Long Double-Stranded RNAs in Plastids', *Science* 347, no. 6225 (2015): 991–94. doi.org/10.1126/science.1261680

22 A.W.E. Franz et al., 'Engineering RNA Interference-Based Resistance to Dengue Virus Type 2 in Genetically Modified *Aedes aegypti*', *Proceedings of the National Academy of Sciences of the United States of America* 103, no. 11: 4198–203. doi.org/10.1073/pnas.0600479103

23 A. Regalado, 'The Next Great GMO Debate', *MIT Technology Review*, 11 August 2015. technologyreview.com/s/540136/the-next-great-gmo-debate/

24 J. Thomas, 'The National Academies' Gene Drive Study Has Ignored Important and Obvious Issues', *Guardian*, 9 June 2016. theguardian.com/science/political-science/2016/jun/09/the-national-academies-gene-drive-study-has-ignored-important-and-obvious-issues

25 T. Nolan and A. Crisanti, 'Using Gene Drives to Limit the Spread of Malaria Introducing Genetic Changes into Mosquito Populations Could be Key to Effective Malaria Control', *The Scientist*, 1 January 2017. the-scientist.com/?articles.view/articleNo/47755/title/Using-Gene-Drives-to-Limit-the-Spread-of-Malaria

26 Ibid.

27 Thomas, 'The National Academies' Gene Drive Study' (2016).

28 A. Regalado, 'Meet the Moralist Policing Gene Drives, a Technology that Messes with Evolution', *MIT Technology Review*, 7 June 2016. technologyreview.com/s/601634/meet-the-moralist-policing-gene-drives-a-technology-that-messes-with-evolution/

29 'A Call for Conservation with a Conscience: No Place for Gene Drives in Conservation', *SynBioWatch*, September 2016. synbiowatch.org/wp-content/uploads/2016/09/letter_vs_genedrives.pdf

30 National Academies of Sciences, Engineering, and Medicine, *Gene Drives on the Horizon: Advancing Science, Navigating Uncertainty, and Aligning Research with Public Values* (Washington, DC: The National Academies Press, 2016), 10. nap.edu/23405

31 P.K. Dearden et al., 'The Potential for the Use of Gene Drives for Pest Control in New Zealand: A Perspective', *Journal of the Royal Society of New Zealand* (2017): 10. doi.org/10.1080/03036758.2017.1385030

32 M.J. Way and K.C. Koo, 'Role of Ants in Pest Management', *Annual Review of Entomology* 37, no. 1 (199): 479–503. doi.org/10.1146/annurev.ento.37.1.479

33 P.J. Lester et al., 'Determining the Origin of Invasions and Demonstrating a Lack of Enemy Release from Microsporidian Pathogens in Common Wasps (*Vespula vulgaris*)',

Diversity and Distributions 20, no. 8 (2014): 964–74. doi.org/10.1111/ddi.12223

34 Q.-H. Fan, Z.-Q. Zhang, R. Brown and S. Bennett, 'New Zealand *Pneumolaelaps* Berlese (Acari: Laelapidae): Description of a New Species, Key to Species and Notes on Biology', *Systematic & Applied Acarology* 21, no. 1 (2016): 119–38. doi.org/10.11158/saa.21.1.8

35 J.K. Dobelmann et al., 'Fitness in Invasive Social Wasps: The Role of Variation in Viral Load, Immune Response, and Paternity in Predicting Nest Size and Reproductive Output', *Oikos* 126, no. 8 (2017): 1208–218. doi.org/10.1111/oik.04117

36 R.D. Akre, 'Wasp Research: Strengths, Weaknesses, and Future Directions', *New Zealand Journal of Zoology* 18, no. 2 (1991): 223–27. doi.org/10.1080/03014223.1991.10757970

37 See Predator Free New Zealand, predatorfreenz.org/

38 C.M. King, 'Failed proposals to Import the Mongoose, Pine Marten, Patagonian Fox and other Exotic Predators into New Zealand', *Journal of the Royal Society of New Zealand* 48, no. 1 (2018): In press. doi.org/10.1080/03036758.2017.1389755

39 Thomas, *The European Wasp* (1960), 60.

40 N.J. Gemmell et al., 'The Trojan Female Technique: A Novel, Effective and Humane Approach for Pest Population Control', *Proceedings of the Royal Society*, Series B 280, no. 1773 (2013): 3. doi.org/10.1098/rspb.2013.2549

41 J. Beggs, 'The Ecological Consequences of Social Wasps (*Vespula* spp.) Invading an Ecosystem that Has an Abundant Carbohydrate Resource', *Biological Conservation* 99, no. 1 (2001): 17–28. doi.org/10.1016/S0006-3207(00)00185-3

6. Judge, jury, executioner?

1 D.I. Theodoropoulos, *Invasion Biology: Critique of a Pseudoscience* (Blythe, CA: Avvar Books, 2003), 185.

2 Ibid., 114.

3 H.S. Rogers et al., 'Effects of an Invasive Predator Cascade to Plants via Mutualism Disruption', *Nature Communications* 8, no. 14557 (2017). doi.org/10.1038/ncomms14557

4 M.A. Davis et al., 'Don't Judge Species on their Origins', *Nature* 474, no. 1 (2011): 153–54. doi.org/10.1038/474153a

5 D. Simberloff, 'Non-Natives: 141 Scientists Object', *Nature* 475, no. 7354 (2011): 36. doi.org/10.1038/475036a

6 P. Salo et al., 'Alien Predators are More Dangerous than Native Predators to Prey Populations', *Proceedings of the Royal Society* Series B 274, no. 1615 (2007): 1237–243.

doi.org/10.1098/rspb.2006.0444

7 Ibid., 1242.

8 J.C. Russell, 'Do Invasive Species Cause Damage? Yes', *BioScience* 62, no. 3 (2012): 217.

9 H.P. Jones et al., 'Invasive-Mammal Eradication on Islands Results in Substantial Conservation Gains', *Proceedings of the National Academy of Sciences U.S.A* 113, no.15 (2016): 4033–038. doi.org/10.1073/pnas.1521179113

10 M.A. Davis, 'Do Invasive Species Cause Damage? Yes. – A Response from Davis', *Bioscience* 62, no. 3 (2012): 217. doi.org/10.1525/bio.2012.62.3.21

11 J.C. Russell, 'A Comparison of Attitudes towards Introduced Wildlife in New Zealand in 1994 and 2012', *Journal of the Royal Society of New Zealand* 44, no. 4 (2014): 136–151. doi.org/10.1080/03036758.2014.944192

12 W. Fraser, 'Introduced Wildlife in New Zealand: A Survey of General Public Views', *Landcare Research Science Series*, no. 23 (2001): 1–46. doi.org/10.7931/DL1-LRSS-23

13 I.L. Neubauer, 'Australia's Aborigines Launch a Bold Legal Push for Independence', *Time Magazine*, 30 May 2013. world.time.com/2013/05/30/australias-aborigines-launch-a-bold-legal-push-for-independence/?iid=gs-main-lead

14 'Read the Treaty: Differences Between the Texts', *NZ History*, accessed September 2017. nzhistory.govt.nz/politics/treaty/read-the-Treaty/differences-between-the-texts

15 A. Gawande, 'The Mistrust of Science', *New Yorker*, 10 June 2016. newyorker.com/news/news-desk/the-mistrust-of-science

16 K.L.E. Kaiser, 'Pesticide Report: The Rise and Fall of Mirex', *Environmental Science and Technology* 12, no. 5 (1978): 520–28. doi.org/10.1021/es60141a005

17 C.J. Robinson and T.J. Wallington, 'Boundary Work: Engaging Knowledge Systems in Co-Management of Feral Animals on Indigenous Lands', *Ecology and Society* 17, no. 2 (2012): 16. doi.org/10.5751/ES-04836-170216

18 M. Kazatchkine, J. Kinderlerer and A. Gilligan, 'Brussels Declaration: Twenty-Point Plan for Science Policy', *Nature* 541, no. 7637: 289. doi.org/10.1038/541289a

7. Where wasps aren't despised (quite so much)

1 From a translation of *Hortus Sanitatis* by Andrewe in 1521, as given in R. Edwards, *Social Wasps: Their Biology and Control* (East Grinstead, UK: Rentokil Ltd., 1980), 17.

2 D. Bolton, 'Angry, Drunk and Unemployed German Wasps are Invading Essex', *Independent*, 25 August 2015. independent.co.uk/news/uk/home-news/angry-drunk-and-

unemployed-german-wasps-are-invading-essex-10471608.html

3 B. Lazarus, 'IT'S THE LUFTWASPE! German "SUPER" Wasps who Sting Repeatedly Threaten to Invade British Gardens and Parks this Summer', *The Sun*, 27 May 2017. thesun.co.uk/news/3665651/german-super-wasps-threaten-invade-parks/

4 R. O'Donoghue, 'Experts Warn of Incredibly Aggressive Drunk German WASPS Plaguing Brits', *Daily Star*, 29 August 2016. dailystar.co.uk/news/latest-news/541630/german-wasps-yellow-jackets-drunk-aggressive-cleankill-environmental-services

5 Exodus 23: ll. 28–30.

6 Joshua 24: l. 12.

7 R. Jamieson, *Jamieson, Fausset, and Brown's Commentary On the Whole Bible* (Michigan: Zondervan, 1999). biblestudytools.com/commentaries/jamieson-fausset-brown/joshua/joshua-24.html

8 T. Branigan, 'Hornet Attacks Kill Dozens in China', *Guardian*, 26 September 2013. theguardian.com/world/2013/sep/26/hornet-attacks-kill-18-china

9 M. Plotkin et al., 'Solar Energy Harvesting in the Epicuticle of the Oriental Hornet' (*Vespa orientalis*)', *Naturwissenschaften* 97, issue 12 (2010): 1067–076. doi.org/10.1007/s00114-010-0728-1

10 J.S. Ishay, 'Hornet Flight is Generated by Solar Energy: UV Irradiation Counteracts Anaesthetic Effects', *Journal of Electron Microscopy* 53, no. 6 (2004): 623–33. doi.org/10.1093/jmicro/dfh077

11 P. Barkham, 'Conservationists Slam "Hateful" Survey Promoting Wasp Killing', *Guardian*, 23 August 2017. theguardian.com/environment/2017/aug/23/conservationists-slam-hateful-survey-promoting-wasp-killing

12 S. Sumner and R. Brock, 'In Defence of Wasps: Why Squashing Them Comes with a Sting in the Tale', *The Conversation*, 12 July 2016. theconversation.com/in-defence-of-wasps-why-squashing-them-comes-with-a-sting-in-the-tale-60729

13 R. Radar, 'Non-Bee Insects are Important Contributors to Global Crop Pollination', *Proceedings of the National Academy of Sciences, USA* 113, no. 1: 146–51. doi.org/10.1073/pnas.1517092112

14 A.C. Hallett, R.J. Mitchell, E.R. Chamberlain and J.D. Karron, 'Pollination Success Following Loss of a Frequent Pollinator: The Role of Compensatory Visitation by Other Effective Pollinators', *AoB Plants* 9, no. 3: 1–11. doi.org/10.1093/aobpla/plx020

15 Sumner and Brock, 'In Defence of Wasps'.

16 R. Edwards, *Social Wasps: Their Biology and Control* (East Grinstead: Rentokil Ltd., 1980), 210.

17 M.W.F. Tweedie, *Poisonous Animals of Malaya* (1941), cited in J. van der Vecht, 'The

Vespinae of the Indo-Malayan and Papuan Areas (Hymenoptera, Vespidae)', *Zoologische Mededeelingen*, 36, no. 13 (1959): 205–32. repository.naturalis.nl/document/148846

18 M.J. Sheehan and E.A. Tibbetts, 'Specialized Face Learning is Associated with Individual Recognition in Paper Wasps', *Science* 334, no. 6060 (2011): 1272–275. doi.org/10.1126/science.1211334

19 E.A. Tibbetts, 'Visual Signals of Individual Identity in the Wasp *Polistes fuscatus*', *Proceedings of the Royal Society* Series B 269, no. 1499 (2002): 1423–428. doi.org/10.1098/rspb.2002.2031

20 A. Avarguès-Weber et al., 'Recognition of Human Face Images by the Free Flying Wasp *Vespula vulgaris*', *Animal Behavior and Cognition* 4, no. 3 (2017): 314–23. doi.org/10.26451/abc.04.03.09.2017

21 F.H. Lawson, R.L. Rabb, F.E. Guthrie and T.G. Bowery, 'Studies of an Integrated Control System for Hornworms on Tobacco', *Journal of Economic Entomology* 75, no. 1 (1961): 93–97. doi.org/10.1093/jee/54.1.93

22 Ibid., 95.

23 N.E. Stamp and M.D. Bowers, 'Indirect Effect on Survivorship of Caterpillars due to Presence of Invertebrate Predators.' *Oecologia* 88, no. 3 (1991): 325–30. doi.org/10.1007/BF00317574

24 W.A. Smirnoff, 'Predators of *Neodiprion swainei* Midd. (Hymenoptera: Tenthredinidae) Larval Vectors of Virus Diseases', *The Canadian Entomologist* 91, no. 4 (1959): 246–48. doi.org/10.4039/Ent91246-4

25 J.R.P. Parra, W. Goncalves, S. Gravena and A.R. Marconato, 'Parasitos e Predadores do Bicho-Mineiro do Caffeeiro *Perileucoptera coffeella* (Guérin-Méneville, 1842) en São Paulo', *Annals of the Society of Entomology.* 6, no. 1 (1977): 138–43.

26 M. Raveret Richter, 'Social Wasp (Hymenoptera: Vespidae) Foraging Behaviour', *Annual Review of Entomology* 45, no. 1 (2000): 121–50. doi.org/10.1146/annurev.ento.45.1.121

27 B.J. Donovan, 'Potential Manageable Exploitation of Social Wasps, *Vespula* spp. (Hymenoptera: Vespidae), as Generalist Predators of Insect Pests', *International Journal of Pest Management* 49, no. 4 (2003): 281. doi.org/10.1080/0967087031000123698

28 R.J. Harris, 'Frequency of Overwintered *Vespula germanica* (Hymenoptera: Vespidae) Colonies in Scrubland-Pasture Habitat and their Impact on Prey', *New Zealand Journal of Zoology* 23, no. 1 (1996): 11–17. doi.org/10.1080/03014223.1996.9518061

29 Donovan, 281.

30 F. Barbero et al., 'Queen Ants Make Distinctive Sounds That Are Mimicked by a Butterfly Social Parasite', *Science* 323, no. 5915 (2009): 782–85.

doi.org/10.1126/science.1163583

31 J.A. Thomas, D.J. Simcox and R.T. Clarke, 'Successful Conservation of a Threatened *Maculinea* Butterfly', *Science* 325, no. 5936 (2009): 80–83. doi.org/10.1126/science.1175726

32 'Blue Butterfly Colonies Thriving', *BBC News*, 16 June 2009. news.bbc.co.uk/1/hi/sci/tech/8102739.stm

33 J.A. Thomas et al., 'Insect Communication: Parasitoid Secretions Provoke Ant Warfare – Subterfuge used by a Rare Wasp may be the Key to an Alternative Type of Pest Control', *Nature* 417, no. 6891 (2002): 505–06. doi.org/10.1038/417505a

34 P. MacIntyre and J. Hellstrom, *An Evaluation of the Costs of Pest Wasps (*Vespula *species) in New Zealand*, Department of Conservation and Ministry for Primary Industries, 1–44.

8. A world with or without invasive wasps?

1 Apsley Cherry-Garrard, *The Worst Journey in the World: Antarctic 1910–1913, vol. 1* (New York, NY: George H. Doran Company; London: Constable and Company, 1922), 64. archive.org/details/worstjourneyinwo01cher

2 S.N. Higgins and M.J. Van der Zanden, 'What a Difference a Species Makes: A Meta-Analysis of Dreissenid Mussel Impacts on Freshwater Ecosystems', *Ecological Monographs* 80 (2010): 179–96. doi.org/10.1890/09-1249.1

3 C.E. Hebert et al., 'Ecological Tracers Track Changes in Bird Diets and Possible Routes of Exposure to Type E Botulism', *Journal of Great Lakes Research* 40 (2014): 64–70. doi.org/10.1016/j.jglr.2013.12.015

4 B. Bienkowski, 'Mass Murder by Botulism: Surge in Great Lakes Bird Deaths Driven by Invaders', *The Georgia Straight*, 18 September 2014. straight.com/news/733061/mass-murder-botulism-invasive-species-drive-bird-death-surge

5 D. Simberloff, 'Nature's Nature and the Place of Non-native Species', *Current Biology* 25 (2015): R585–R599.

6 Convention on Biological Diversity, 'Aichi Targets', accessed October 2017. cbd.int/sp/targets/

7 SDSN, 'Indicators and a Monitoring Framework', accessed October 2017. indicators.report/targets/15-8/

8 D.P. Tittensor et al. 'A Mid-term Analysis of Progress Toward International Biodiversity Targets', *Science* 346 (2014): 241–44. doi.org/10.1126/science.1257484

9 IUCN, 'The Honolulu Challenge on Invasive Alien Species', accessed October 2017. iucn.org/theme/species/our-work/invasive-species/honolulu-challenge-invasive-alien-species

10 F. Pearce, 'Tackling Britain's "Green Xenophobia" over Alien Plants and Animals: Our Ecological Systems can be Helped, not Harmed, by Incoming Flora and Fauna', *Independent*, 13 April 2015. independent.co.uk/environment/nature/tackling-britains-green-xenophobia-over-alien-plants-and-animals-our-ecological-systems-can-be-10174455.html

11 F. Pearce, *The New Wild: Why Invasive Species Will Be Nature's Salvation* (Beacon Press, Boston, 2015), 106.

12 Ibid., 107.

13 Biosecurity Act 1993 No 95, accessed October 2017. legislation.govt.nz/act/public/1993/0095/155.0/DLM314623.html

14 Greater Wellington Regional Pest Management Strategy 2002–2022, Five Year Review 2007, accessed October 2017. gw.govt.nz/assets/Our-Environment/Biosecurity/Pest-plants/GreaterWellingtonRPMS2009.pdf

15 S. Connor, 'Rare Birds Return to Remote South Georgia Island after Successful Rat Eradication Programme', *Independent*, 25 June 2015. independent.co.uk/environment/nature/rare-birds-return-to-remote-south-georgia-island-after-successful-rat-eradication-programme-10345864.html

16 R.W. Elwood and L. Adams, 'Electric Shock Causes Physiological Stress Responses in Shore Crabs, Consistent with Prediction of Pain', *Biology Letters* 11 (2015): 20150800. doi.org/10.1098/rsbl.2015.0800

17 D. Santoro, S. Hartley, D.M. Suckling and P.J. Lester, 'The Stinging Response of the Common Wasp (*Vespula vulgaris*): Plasticity and Variation in Individual Aggressiveness', *Insectes Sociaux* 62 (2015): 455–63. doi.org/10.1007/s00040-015-0424-4

18 Joe Ballenger, 'Do Insects Feel Pain?', *Ask an Entomologist* (blog), 29 August 2016. askentomologists.com/2016/08/29/do-insects-feel-pain/

INDEX